BEFORE IT'S TOO LATE

TOO LATE

A Scientist's Case
for Nuclear Energy

BEFORE IT'S TOO LATE
A Scientist's Case for Nuclear Energy

BERNARD L. COHEN

PLENUM PRESS • NEW YORK AND LONDON

Library of Congress Cataloging in Publication Data

Cohen, Bernard Leonard, 1924–

    Before it's too late.

    Includes bibliographical references and index.
    1. Atomic power. I. Title.
TK9145.C576 1983   333.79′24   83-11083
ISBN 0-306-41425-2

FOREWORD

I was not invited to write a foreword for this book. Dr. Cohen, knowing my busy schedule, would have considered such a request to be an imposition. I volunteered to do so in part to acknowledge my gratitude to him for having been a constant source of reference materials as I have turned my attention increasingly to informing both lay and scientific audiences concerning the biologic effects of low-level ionizing radiation. My primary reason for volunteering, however, is to point to the importance of such a book for public education at a time when the media, in collaboration with a variety of activist groups, have developed among the people an almost phobic fear of radiation at any level.

I take issue with the words of another Nobel laureate, George Wald, who states regularly "Every dose is an overdose."[1] This philosophy has resulted in women refusing mammography for the detection of breast cancer even though this methodology is the most sensitive for detection of such cancers in the early, curable stage, and even though, at present, breast cancer is the leading cause of cancer deaths among women. It has led a Westchester County, New York legislator to state proudly in the *New York Times* that he had introduced legislation that would bar *all radioactivity* from the county's roads. It is indeed a tragic state of affairs when elected officials are so ignorant as not to know that all living things are radioactive, and that the building materials for roads are radioactive as well. Had that legislation passed there would have been no need for roads—no one could have used them! An increased understanding of the concept of *"de minimus,"* an amount of radioactivity or a radiation dose so low that any conceivable deleterious effects would be negligible, is essential if science instead of mysticism is to govern our lives.

In this book Dr. Cohen has attempted to familiarize the reader with radiation-related risks as compared to other risks in daily living. In the interests of conservatism he has adapted the "linear extrapolation hypothesis," which accepts that one can extrapolate from deleterious effects observed at high dose rates to predict the probability of similar effects at low doses and dose rates. I believe that this book fails to emphasize sufficiently that this hypothesis consistently exaggerates the possible effects at low doses and dose rates. There are numerous papers demonstrating that for a given cumulative dose, cancer induction is dose-rate related.[2] Let me give one example from the book of the conservatism of Dr. Cohen's approach: In Chapter 6, in dealing with the radon problem, Dr. Cohen uses the data on uranium miners, applies the linear extrapolation hypothesis, and estimates that environmental radon exposure is now causing about 10,000 fatal lung cancers each year in the United States. If we use the age-adjusted cancer death rates (age distribution of the 1940 U.S. Census population) for lung cancer for men (2.6/100,000) and women (1.6/100,000) for 1930 in place of the 1976 rates similarly age-adjusted (53/100,000 and 16/100,000) one would expect 3600 lung cancer deaths among men and 2500 among women.[3] Thus, if smoking were eliminated as a factor there would have been only 6000 lung cancer deaths, rather than the 100,000 estimated for 1979. The lung cancer death rate in 1930 was likely not all attributable to radon, since even at that time some lung cancer deaths were due to smoking or other carcinogenic pollutants and some were spontaneous. Thus the 10,000 estimated radon-related cancers are overly conservative. In fact, even the trivial radiation effects described by Dr. Cohen are probably much higher than the true effects—which may well be zero.

This is an important book. Regrettably, I doubt it will be read by the antinuclear activists who need it the most, or by those possessed by a phobic fear of radiation. Our nation survived the Salem witch trials. It is difficult for one like me, trained in a scientific discipline, to understand a nuclear phobia that appears to be derived from a belief in the unknown and the mysterious resembling the hysteria that was acceptable 300 years ago. If the

American people are to remain proud of our standard of living and our ability to help the less privileged in the world, we must develop scientific literacy among all our people. Those who make our laws are subject to pressures from their constituents. It is important that the pressures come from an informed electorate. This book is a step in the right direction.

Rosalyn S. Yalow

REFERENCES

1. E. Sternglass, *Secret fallout: Low-level radiation from Hiroshima to Three Mile Island* (McGraw-Hill, New York, 1981).
2. National Council on Radiation Protection and Measurements (NCRP), "Influence of dose and its distribution in time on dose–response relationships for low-level radiations," NCRP Report No. 64 (1980).
3. L. Garfinkel, and E. Silverberg, *Cancer Statistics,* 1979 (American Cancer Society, New York, 1979).

PREFACE

There have been several books about nuclear power over the past few years. Why write another one?

Nuclear power is widely viewed as a political issue, but it certainly has a large scientific and technical component, which must be understood before sound political judgments can be made. Most of the books have been written by journalists and other professional writers with little or no training in science and technology. There is nothing wrong with that. Whereas a scientist's job is doing research to develop information, a journalist's job is getting the information from him to present to the public in understandable terms. This system has generally worked reasonably well, as, for example, in our exploration of outer space, or in the wonderful progress of medical science over the last several decades. It is an efficient system, for science is a very demanding discipline and scientists are typically deeply engrossed in its details. They generally have little talent for or interest in writing for the public, having little time or opportunity to practice and develop the skills needed for that purpose. Why not give their information to professional writers with their well-developed skills to present it to the public?

Somehow, in the case of nuclear power, that system has failed, and failed miserably. The journalists have *not* transmitted the information. Scientists are accustomed to inaccuracy, incompleteness, and misplaced emphasis in journalistic coverage of science. But in this case, the failure in transmission has gone far beyond that. The information has been so hopelessly garbled that the overall sense of the message transmitted is just plain wrong. Numerous examples of this will be cited throughout this book.

The reasons for this complete breakdown in transmission of information from the scientific community to the public via journalism are difficult to fathom. I have been interviewed on several occasions by journalists and professional writers, who just do not seem to understand the basic points. They take copious notes on all sorts of trivia, but not on the important things I have to say. I get the impression that they are looking for things that will frighten the public. They seem to want human interest stories, but to be turned off by the results of lengthy and complex quantitative analyses. Unfortunately, the latter are the business of science, while the former have no scientific value. Such stories may be useful for attracting an audience, but they are hardly a basis for drawing broad conclusions.

For whatever reason, the scientific and technical information the public requires to make intelligent political decisions on nuclear power has not been transmitted by journalists and professional writers. On the contrary, the information that has been transmitted is highly deceptive and is basically false.

A few scientists have written popular books about nuclear power issues, but they have generally been "fringe types" who have largely abandoned their scientific careers. They spend most of their time speaking and writing to present viewpoints that are contrary to those accepted by the scientific community. I hesitate to speculate on their motives because my ideas on that subject are highly uncharitable. But the important point is that they do *not* represent the mainstream of thought in the scientific community. Nevertheless, they have succeeded in getting their distorted message to journalists, and the latter have amplified the distortions in transmitting it to the public.

As the situation stands, the mainstream of the scientific community has not been heard on this subject. Worse than that, it has been maligned and its credibility questioned. We nuclear scientists have come to feel isolated and under attack. We are not accustomed to public combat and are ill-prepared to engage in it, but I have concluded that engage in it we must.

If our democracy is to work, and the public is to make decisions on issues that have important scientific components, it must get information from the scientific community. In the case of nuclear power the traditional lines of information transmittal have failed miserably. Other approaches must be undertaken. After all, nuclear power is a vital national issue, involving a trillion dollars of the public's money; it may well determine whether our nation enters the 21st century flourishing in prosperity or decaying in poverty.

That is the reason I decided to write this book—for once, to get the viewpoint of the main-line scientific community across to the public. My 35 years of 60-hour weeks working in the laboratory have done little to prepare me for this task. I can only hope that the reader will forgive any lack of communication skills. My editor, Linda Regan, has been most helpful in that regard, helping me immensely. She has taught me not to split infinitives, dangle participles, or run sentences on and on. She has also provided a great many important suggestions about content. Her patient reading of draft after draft of this manuscript is gratefully acknowledged.

I am also indebted to innumerable scientific colleagues from all over the country who have provided me with information, offered suggestions, and encouraged my efforts. I could not possibly begin to name them all, but their collective contribution has been enormous. Last but not least, I want to acknowledge the assistance of my secretary, Mary Gromicko, whose infinite patience in typing and retyping, checking spelling, and many other tasks has been invaluable. I only hope that our combined efforts succeed in giving you, the reader, the knowledge and understanding you need to participate intelligently in the vital decision-making process about nuclear power.

Bernard L. Cohen

CONTENTS

BEFORE IT'S
TOO LATE

Chapter 1 / PROBLEMS IN PUBLIC UNDERSTANDING

How well does the American public understand the hazards of nuclear power? A poll of radiation health scientists* shows that 82% of them feel that the public's fear of radiation is "substantially" or "grossly" exaggerated. Another poll* shows that 89% of all scientists, and 95% of all scientists involved in energy-related fields, favor proceeding with the development of nuclear power; among the public there is only a slight majority in favor.

In a recent study in Oregon, groups of college students and members of the League of Women Voters were asked to rank 30 technologies and activities according to the "present risk of death" they pose to the average American.[1] Both groups ranked nuclear power No. 1, well ahead of motor vehicles, which kill about 50,000 Americans each year, cigarette smoking, which kills 150,000, and eleven others that each kill over 1,000. How many can be expected to die annually from generation of nuclear power including the risk of accidents, radioactive waste, and all of the other dangers we hear so much about? According to estimates developed by government-sponsored research pro-

* Cf. Chapter 10.

1

grams, about *ten* per year.[2] If you don't trust "the Establishment," you might trust the leading antinuclear activist organization in the United States, the Union of Concerned Scientists (UCS), which estimates an average of 120 deaths per year[3] from nuclear power. In either case it is obvious that nuclear power is perceived to be *thousands of times* more dangerous than it actually is, even by college students and League of Women Voters members who are well above average in intelligence, education, and awareness of public issues. Clearly the American public is grossly misinformed about the hazards of nuclear power.

A poll by the Opinion Research Corporation[4] found that about 80% of the American public believes that it is more dangerous to generate electricity from nuclear power than by burning coal. The relative danger of nuclear and coal-burning electricity generation is basically a scientific question; it has been dealt with by many scientific studies. All of them agree that, contrary to public opinion, coal is the more dangerous.[5] This includes studies sponsored by the U.S. National Academy of Sciences, the American Medical Association, the United Kingdom Health and Safety Executive, the Norwegian Ministry of Oil and Energy, the State Legislatures of Maryland and of Michigan, and many others. It has been conceded by the antinuclear activist UCS,[3] by its Director Henry Kendall,[6] and privately even by Mr. Anti-Nuke, Ralph Nader.[7] I know of no study that has reached the opposite conclusion, that nuclear energy generation is more dangerous than coal. In a 1981 magazine article,[8] I offered a $50 reward for information leading to my discovery of such a study, but in all this time no one has tried to claim that reward. Clearly the 80% of the American public that believes coal burning to be safer than nuclear energy is *grossly misinformed*. Their numbers, incidentally, seem to include Walter Cronkite, who stated in a 1979 broadcast "nuclear energy is too dangerous—we must turn to coal instead."

A recent poll[9] shows that 56% of the American public is opposed to having a nuclear power plant in their own community, whereas another[10] shows that this opinion is shared by only 31% of all scientists, by only 20% of scientists specializing in fields related to energy, and by only 2% of scientists specializing in radiation or nuclear science.

Why are there such vast differences between the dangers of nuclear power perceived by the public and those understood by scientific experts? Basically because they are derived from different sources. The public derives its perceptions from the news media, while involved scientists depend on the *scientific literature*.

The scientific literature consists principally of many hundreds of periodicals, each covering a specialized area of science. In them, scientists present research results to their colleagues in sufficient quantitative detail to allow them to be thoroughly understood and checked. They also contain critiques from researchers who may disagree with the procedures used, and replies to these critiques from the original authors, although only a few percent of the papers published are sufficiently controversial to draw such criticism. The whole system is set up to maximize exchange of information and give full airing of the facts. On the other hand, most people would not consider the scientific literature to be interesting reading. It is written by research scientists—not usually skillful writers—to be read by other scientists in the same field. It makes extensive use of mathematics and specialized vocabulary, with little attention to techniques for holding the reader's attention. That function is provided by the scientist's professional need to obtain the information.

News media material, on the other hand, is prepared, in general, by people with little understanding of science, or even of the background necessary to learn about it. They report for a high-school-level audience, and holding its attention is a primary concern. With these limitations and needs, the media are forever looking for political and emotional human interest angles which make for more interesting reading; they generally avoid quantitative scientific analysis because it is not usually interesting or understandable to most readers or even to the writers themselves.

In view of these vast differences between the sources of information used by involved scientists and the general public, it is not difficult to understand how they can hold very different opinions. Surely there can be no question but that those of the scientists are more reliable. But it is public opinion that controls decision making in a democracy like ours.

The power of this public opinion is difficult to overestimate: A few years ago I did a study of how much money our government is willing, or just barely unwilling, to spend to save the life of one of its citizens.[12] I identified many ways in which thousands of American lives could be saved each year by cancer screening or medical treatment programs, or by installation of highway safety equipment, at an average cost of less than $100,000 per life saved. At the other end of the scale, our government is spending over $100 million per life saved on programs for protecting the public from nuclear wastes. Clearly, the public could be protected much better and billions of dollars could be saved if this latter money were spent on medical or highway safety programs. How did this irrational situation come about? In order to find out one need only query the government officials who make these decisions. The answer is very simple: the first priority in our government is to be responsive to public concern. If the public is concerned about radiation, our government will spend tremendous sums to protect it from radiation; but if the public is not particularly concerned about highway safety, spending in that area will be meager. The fact that thousands of lives are lost unnecessarily each year in this allocation of expenditures is of secondary importance.

If one finds this objectionable, one need only ask what would happen to a government official who refused to be responsive to public concern. How long would he last in office? Responsiveness to public concern is a vital part of our democracy and we would not want it to be otherwise. The problem is not in the workings of our government, but rather in misguided public opinion.

There are many other ways in which numerous lives are unnecessarily lost due to the public's highly exaggerated fear of nuclear power. This fear is driving utilities to build coal-fired instead of nuclear power plants. Every time this happens—typically 10 times each year—many hundreds of citizens are unnecessarily condemned to an early death. It will be shown in Chapter 4 that this conclusion is valid even if we use the estimates offered by the antinuclear activists.

But probably more important is the drain public fear of nuclear power imposes on our wealth. We will show in Chapter 4 that wealth creates health, and that this drain can easily cause tens of thousands of American deaths each year. With these large numbers of unnecessary deaths, our story is beginning to sound like a horrible tragedy, but more is yet to come.

Quite aside from health issues, two of the most serious environmental problems we face are air pollution and acid rain. Air pollution is doing billions of dollars worth of damage each year, spreading filth and ugliness, and destroying a wide variety of property ranging from women's stockings to granite statues. Acid rain is rendering lakes lifeless, damaging the forestry and fishing industries, and creating international tensions between the United States and Canada. Both the air pollution and acid rain problems could be largely eliminated by large-scale use of nuclear power. The tragedy of the misunderstanding deepens.

As we project into the future, the tragedy multiplies. Burning coal, oil, and gas is causing earthshaking climatic changes that could eventually turn our Midwestern grain belt into a desert, and flood out our coastal cities— New York, Miami, New Orleans, Houston, Los Angeles, and a host of others. Nuclear power could prevent this if the misunderstandings about its dangers could be eliminated.

The most important problem for our distant progeny will be a shortage of materials that we now obtain by mining. We are now consuming the world's scarce mineral resources at a voracious rate; indeed our era has been called "the age of mining," because within less than a century there will be very little left to mine—no copper, no tin, no lead, no mercury, no zinc, and so on. In the desperate search for substitutes, the most fruitful source would be plastics and organic chemicals. But these are made from coal, oil, and gas, which we are now simply burning up at a rate of millions of tons each every day. Wouldn't it be much better if we instead burned uranium, which has no other important uses, leaving the coal, oil, and gas for future generations to use as a source of materials they will so sorely need? The only thing preventing this is the public's misconception of the dangers of nuclear power.

It would be easy to go on about the benefits of nuclear power to future generations, such as how it could provide all the energy mankind will ever need without price increases due to fuel scarcity. But I hope the misunderstandings will be straightened out in time to avoid tragedy that far in the future.

To return to the present, perhaps the greatest tragedy of all due to this misunderstanding is in the loss of our nation's opportunity to become energy independent. If our nuclear power program had not been sidetracked, we could have achieved energy independence well before the end of this century. The economic benefits would have been enormous, but probably the greatest benefit would have been world peace. If one tries to imagine scenarios that can lead to a large scale nuclear war, high on the list would be fighting over Middle East oil, or oil and gas from several other areas of the world. This danger could have been avoided by maintaining our nuclear power program. With a reasonable number of electric automobiles and heat pumps for warming buildings, oil would hardly be an issue in world politics.

All of these matters together add up to what is probably the greatest tragedy in American history. They will all be discussed in greater detail in later chapters, but our first order of business must be to clear up the public misconceptions that have led to this tragedy. As I see it, the most important problems in public understanding of nuclear power are the following:

1. A wildly exaggerated fear of radiation
2. A highly distorted picture of a reactor meltdown accident
3. A failure to understand and quantify risk
4. The grossly unjustified fears about disposal of radioactive waste
5. The falsely imagined connections between nuclear power and nuclear weapons
6. The romantic idea that solar electricity could or should replace nuclear or coal

Before launching into these subjects, which will be elucidated at length in subsequent chapters, it is perhaps useful to discuss the sources of the information to be presented. These sources are those generally accepted in the scientific community. It may come as a surprise to some readers that nearly all the important facts on these issues are generally accepted (within a degree of uncertainty small enough for the differences of opinion to be of no concern to the public), since the media often given the impression that there are large and important areas of disagreement within the scientific community. Actually, in spite of attempts by outsiders to dramatize it, long-standing controversy is rather rare in science. This is not to say that different scientists don't have different ideas, but rather that there are universally accepted ways of settling disagreements.

We have already mentioned the scientific literature as a mechanism for exposing research results to criticism. In addition, specialists in each field frequently get together at meetings where there is ample opportunity for airing out disagreement before an audience of scientific peers. After these discussions participants and third parties often return to their laboratories to do further measurements or calculations, developing further evidence. In most cases, controversies are thereby settled in a matter of months, leading to a consensus with which over 90% of those involved would agree. Where scientific questions have an impact on public policy, there is an additional mechanism. The National Academy of Sciences and similar national and international agencies assemble committees of distinguished scientists involved in the field to develop and document a consensus. Only very rarely do these committees have a minority report, and then it's from a very tiny minority. The committee's conclusions are generally accepted by scientists and government agencies all over the world. A clear demonstration of this acceptance is given in Chapter 10. Since controversy is the rule rather than the exception in human affairs, many find it difficult to believe that science is so different in this respect. The reason for the difference is that science deals largely with quantities that can be measured and calculated, and these measurements and calculations can be repeated and checked by doubters, with a very heavy professional penalty to be paid by anyone reporting erroneous results. This ability rapidly to resolve controversy has been one of the most important elements in the great success of science, a success that during this century has increased our life expectancy by 20 years, improving our standard of living immeasurably.

When scientists present information to the public, they can easily convey false impressions without falsifying facts, by merely selecting the facts they present. It is my pledge not to do this. I will do my utmost not only to present correct information, but to present it in a way that gives the correct impression and perspective. Since your faith in this pledge may depend on what you know about the author, I offer the following personal information. I am a 58-year-old, long tenured University Physics Professor with no connection to the nuclear industry. While I do some consulting for it, as is necessary to develop and maintain my professional competence, I have always been pledged to use the income I thereby derive (only a few percent of my university salary) only to purchase documents and to finance travel to scientific meetings. (Lest the latter be suspicious, I travel more than I would like.) My job security and salary are in no way dependent on the health of the nuclear industry. I have no long-standing emotional ties to nuclear energy, not having participated in its development; most of my research career was in basic nuclear structure physics from which nuclear power development had split off before I entered the field. My professional involvement with nuclear energy began only when the 1973 oil embargo stimulated me to look into our national energy problems. I have four children and three grandchildren; my principal concern in life is to increase the chances for them and all of our younger citizens to live healthy, prosperous lives in a peaceful world.

To those who question my selection of topics or my treatment of them in this book, I invite personal correspondence or telephone calls to discuss these questions. I feel confident that through such means I can convince any reasonable person that the viewpoints expressed are correct and sufficiently complete to give the proper impressions and perspective.

Other scientists have recently written books on nuclear energy, painting a very different picture from the one I present. Ernest Sternglass,[11] John Gofman,[12] and Helen Caldicott[13] are the names with which I am familiar. Their basic claim is that radiation is far more dangerous than estimates by the scientific "establishment" would lead us to believe it is. This is a scientific question which will be discussed in some detail in Chapter 2, but the ultimate judgment is surely best made by the community of radiation health scientists. A poll of that community (cf. Chapter 10) shows that the scientific works of these three have very low credibility in it. Their ideas on the dangers of radiation have been unanimously rejected by various committees of eminent scientists assembled to make scientific judgments on those questions. These committees represent what might be called "The Establishment" in radiation health science; the poll shows that they have very high credibility within the involved scientific community. Less than 1%, in a secret ballot, gave these "Establishment" groups a credibility rating below 50 on a scale of 0–100, whereas 83% gave the three above-mentioned authors a credibility rating in that low range. The average credibility rating of these Establishment groups was 84, whereas less than 3% of respondents rated the three authors that high.

The positions presented in this book are those of "The Establishment." In comparing this book with those of Sternglass, Gofman, and Caldicott, you therefore should not consider it as my word against that of those three authors, but rather as their positions versus the positions of "The Establishment," strongly supported by a poll taken anonymously of the involved scientific community.

Another approach to judging between this book and theirs is the degree to which the authors are actively engaged in scientific research. The Institute for Scientific Information, based in Philadelphia, keeps records on all papers in scientific journals and publishes listings periodically in its *Science Citation Index—Sourcebook*. The number of publications they list for the various authors, including myself, each year are:*

Author	1975–79	1980	1981	1982 (January–June)
H. Caldicott	3	0	0	1
J. W. Gofman	2	1	2	0
E. J. Sternglass	5	1	1	0
B. L. Cohen	65	11	14	8

It might be noted that most of the papers by Gofman were his replies to critiques of his work, and most of the entries for Sternglass were papers at meetings which are not refereed (the others were on topics unrelated to nuclear power). Only a small fraction of my papers are in these categories.

Of course, there have been many books about nuclear power written by professional writers and other nonscientists. They are full of stories that make nuclear power seem dangerous. But stories are useful principally to maintain reader interest—they don't prove anything. In order to make a judgment on the hazards of nuclear power, it is necessary to quantify the number of deaths (or other health impacts) it can be expected to cause, and compare this with similar estimates for other technologies. This is something that the books by nonscientists never do. They sometimes quantify what "might" happen, but never quantify the *probability* that it will happen. If we are to be guided by what *might* happen, I could easily concoct scenarios for any technology that would result in more devastation and death than any of their stories about nuclear power.

A book of this type must often get into discussions of scientific details. Every effort has been made to keep them as readable as possible. The more technical details have been put into Appendixes at the ends of chapters. They can be ignored by the reader with less interest in details. For readers with

---

* In addition the first three have each published a book, but the scientific research process is not assisted by books written for public consumption as were theirs (and this book).

more interest in these, numerous references are given which can be used as starting points for further reading.

Each chapter is broken up into sections. If a reader is not interested in the subject of a particular section, or if the reader finds it to be too technical, it can usually be skipped over without loss of continuity.

With these preliminaries out of the way, let us proceed to discuss the most important source of misunderstanding, the public's greatly exaggerated fear of radiation.

REFERENCE NOTES

1. P. Slovic, B. Fischoff, and S. Lichtenstein, "Rating the Risks," *Environment*, **21**, 14, April (1979).
2. Effects during our lifetime are summarized in Chapter 4. Longer-term effects are summarized in Chapter 6.
3. Union of Concerned Scientists, "The Risks of Nuclear Power Reactors," Cambridge, Massachusetts (1977). They give 2.4 deaths/GWe-yr, which multiplied by 50 GWe, the total amount generated in the U.S., gives 120 deaths/yr.
4. Opinion Research Corp, "Public Attitudes toward Nuclear Power vs. Other Energy Sources. *ORC Public Opinion Index*, Vol. **38**(17), September (1980).
5. C. L. Comar and L. A. Sagan, "Health Effects of Energy Production and Conversion," *Annual Review of Energy*, **1**, 581 (1976); L. B. Lave and L. C. Freeburg, "Health Effects of Electricity Generation from Coal, Oil, and Nuclear Fuel," *Nuclear Safety*, **14**(5), 409, (1973); S. M. Barrager, B. R. Judd, and D. W. North, "The Economic and Social Costs of Coal and Nuclear Electric Generation," Stanford Research Institute Report, March (1976); Nuclear Energy Policy Study Group, "Nuclear Power—Issues and Choices" (Ballinger, Cambridge, Massachusetts, 1977); National Academy of Sciences Committee on Nuclear and Alternative Energy Systems, "Energy in Transition, 1985–2010" (W. H. Freeman and Co., San Francisco, 1980); American Medical Association Council on Scientific Affairs, "Health Evaluation of Energy Generating Sources," *Journal of the American Medical Association*, **240**, 2193 (1978); H. Inhaber, "Risk of Energy Production," Atomic Energy Control Board Report AECB-1119, Ottawa (1978); R. L. Gotchy, "Health Effects Attributable to Coal and Nuclear Fuel Cycle Alternatives," U.S. Nuclear Regulatory Commission Document NUREG-0332 (1977); D. J. Rose, P. W. Walsh, and L. L. Leskovjan, "Nuclear Power—Compared to What?" *American Scientist*, **64**, 291 (1976); Union of Concerned Scientists, "The Risks of Nuclear Power Reactors," H. Kendall (Director), Cambridge, Massachusetts (1977); Science Advisory Office, State of Maryland, "Coal and Nuclear Power," (1980); Norwegian Ministry of Oil and Energy, "Nuclear Power and Safety," (1978); Ohio River Basin Energy Study (EPA), "Impacts on Human Health from Coal and Nuclear Fuel Cycles," July (1980); United Kingdom

Health and Safety Executive, "Comparative Risks of Electricity Production Systems," (1980); Maryland Power Plant Siting Program, "Power Plant Cumulative Environmental Impact Report," PPSP-CEIR-1 (1975); W. Ramsay, *Unpaid Costs of Electrical Energy* (Johns Hopkins Univ. Press, Baltimore, Maryland, 1979); Legislative Office of Science Advisor, State of Michigan, "Coal and Nuclear Power" (1980); B. L. Cohen, *American Scientist,* **64,** 291 (1976); R. Wilson and W. J. Jones, *Energy, Ecology, and the Environment* (Academic Press, New York, 1974); H. Fischer *et al.,* "Comparative Effects of Different Energy Technologies," Brookhaven National Lab Report BNL 51491 (Sept. 1981); D. K. Myers and H. B. Newcombe, "Health Effects of Energy Development," Atomic Energy of Canada Ltd. Report AECL-6678 (1980).

6. H. Kendall, Physics Colloquium at Carnegie-Mellon University, 1980.
7. In answer to my question, Nader replied, "Maybe we can clean up coal, or maybe we shouldn't burn coal either."
8. B. L. Cohen, "How Dangerous is Radiation?" *Ascent Magazine,* **2,**(4), 9 (1981).
9. R. Kasperson, G. Berk, A. Sharaf, D. Pijawka, and J. Wood, "Public Opinion and Nuclear Energy: Retrospect and Prospect," *Science, Technology, and Human Values,* 11 Spring (1980).
10. S. Rothman and S. R. Lichter, "The Nuclear Energy Debate: Scientists, The Media, and the Public," *Public Opinion,* August/September (1982), p. 47.
11. E. Sternglass, *Secret Fallout: Low Level Radiation from Hiroshima to Three Mile Island* (McGraw Hill, New York, 1981).
12. J. W. Gofman, *Radiation and Human Health* (Sierra Club Press, San Francisco, 1981).
13. H. Caldicott, *Nuclear Madness* (Bantam, New York, 1981).

Chapter 2 / HOW DANGEROUS IS RADIATION?

The most important breakdown in public understanding of nuclear power is in its concept of the dangers of radiation. What is radiation, and how dangerous is it?

Radiation consists of several types of subatomic particles, called gamma rays, neutrons, electrons, and so on, that shoot through space at very high speeds, something like 100,000 miles per *second*. They can easily penetrate deep inside the human body, doing damage to the biological cells of which the body is composed. This damage can cause a fatal cancer to develop, or if it occurs in reproductive cells, it can cause genetic defects in later generations of offspring. When explained in this way, the dangers of radiation seem to be very grave, and for a person to be struck by a particle of radiation appears to be an extremely serious event. So it would seem from the following description in what has probably been the most influential book from the antinuclear movement[1]:

> When one of these particles or rays goes crashing through some material, it collides violently with atoms or molecules along the way. . . . In the delicately balanced economy of the cell, this sudden disruption can

> be disastrous. The individual cell may die; it may recover. But if it does
> recover . . . after the passage of weeks, months or years, it may begin
> to proliferate wildly in the uncontrolled growth we call cancer.

But before we shed too many tears for the poor fellow who was struck by one of these particles of radiation, it should be pointed out that every person in the world is struck by about 15,000 of these particles of radiation every second of his or her life,[2] and this is true for every person who has ever lived and for every person who ever will live. These particles, totalling 500 billion per year, or 40 trillion in a lifetime, are from natural sources, but in addition our technology has introduced new sources of radiation. By far the most important of these is medical X-rays—a typical X-ray bombards us with trillions of particles of radiation.[2]

With all of this radiation exposure, how come we're not all dying of cancer? The answer to that question is *not* that it takes a very large number of these particles to cause a cancer; as far as we know, every single one of them has that potential; as we are frequently told, "no level of radiation is perfectly safe." What saves us, rather, is that the probability for one of these particles to cause cancer is very low,[2] about one chance in 50 quadrillion (50 million billion, or 50,000,000,000,000,000)! Every time a particle of radiation strikes us, we engage in a fatal game of chance at those odds. Of course this is not unique to radiation; we are engaged in innumerable similar games of chance involving chemical, physical, and biological processes that may lead to any form of human malady, and the one involving radiation has odds much more favorable to us than most. Only one-half of one percent* of human cancers are caused by the 40 trillion particles of radiation that hit us over a lifetime,[2] while the other 99.5% are from losing in one of these other games of chance.

Of course every extra particle that strikes us increases our cancer risk, so many people feel that they should go to great lengths to avoid extra radiation. If that is your attitude, there are many things you can do. You can reduce it 10% by living in a wood house rather than a brick or stone house,[3] because brick and stone contain more radioactive materials like uranium, thorium, and potassium. You could reduce it[4] 20% by building a thick lead shield around your bed to reduce the number of hits while you sleep, or you could cut it in half by wearing clothing lined with lead like the cover dentists drape over you when they take X-rays.

But most people don't do these things. Rather, they recognize that life is full of risks. Every time you take a bite of food, it may have a chemical

---

* This estimate does not include the effects of radon, to be discussed later.

that will cause cancer, but still people go on eating, more than necessary in most cases. Every ride or walk we take could end in a fatal accident, but that doesn't keep us from riding or walking. Similarly, the sensible attitude most of us take is not to worry about a little extra radiation; after all, one chance in 50 quadrillion is pretty good odds!

What this discussion should teach us is that hazards of radiation must be treated *quantitatively*. If we stick to qualitative reasoning alone, it is easy to show that nuclear power is bad—it leads to radiation exposure which can cause cancer, which is bad. The trouble with this is that, by a similar type of qualitative reasoning, just about anything else we do can be shown to be harmful: coal or oil burning causes air pollution which kills people, so coal or oil burning is bad; using natural gas leads to explosions which kill people, so burning gas is bad; etc. Any discussion of dangers from radiation must be *quantitative;* otherwise, it can be as completely deceptive as the quote above about the tragedy of being struck by a single particle of radiation. But how often do stories in the news media about radiation include numbers? I always look for them and hardly ever find them.

MEET THE MILLIREM

In order to discuss radiation exposure quantitatively, we must introduce the unit in which it is measured, called the *millirem,** abbreviated mrem. One millirem of exposure corresponds to being struck by approximately seven billion particles of radiation,[2] but it takes into account variations in health risks with particle type and size of person. For example, a large adult and a small child standing side by side in a field of radiation would suffer roughly the same cancer risk and hence would receive the same dose in millirems, although the adult would be struck by more particles of radiation because the adult is a larger target. In nearly all of our discussions we will be considering doses below about 10,000 mrem, which is commonly referred to as "low-level" radiation.

The news media frequently carry stories about incidents in which the public is exposed to radiation; radioactive material falling off a truck; contaminated water leaking out of a tank or seeping out of a waste burial ground; a radioactive source being temporarily misplaced; malfunctions in nuclear plants leading to releases of radioactivity, and so on. Perhaps a hundred of

---

* Other units sometimes used are millirad and milliroentgen. For gamma rays, X-rays, and beta rays, all three of these are essentially the same. The *rem* (1 rem = 1000 millirem), *microrem* (1 millirem = 1000 microrem), etc. are also commonly used.

these stories over the past 35 years have received national television and press coverage. The thing I always look for in these stories is the radiation exposure in millirems, but it is hardly ever given. Eventually it appears in a technical journal, or I trace it down by calls to health officials. On a very few occasions it has been as high as 5–10 mrem, but in the great majority of cases it has been less than 1 mrem. In the Three Mile Island accident, average exposures in the surrounding area[5] were 1.2 mrem—this drew the one-word banner headline "RADIATION" in a Boston newspaper. In the supposed leaks of radioactivity from a low-level waste burial ground near Moorhead, Kentucky, there were no exposures[6] as high as 0.1 mrem; yet this was the subject of a three-part series in a Philadelphia newspaper[7] bearing headlines "It's Spilling All Over the U.S.," "Nuclear Grave is Haunting KY," and "There's No Place to Hide." In the highly publicized leak from a nuclear power plant near Rochester, New York, in 1982, no member of the public was exposed[8] to as much as 0.3 mrem. Yet this was the top news story on TV and in newspapers for more than 24 hours.

For purposes of discussion, let us say that a typical exposure in these highly publicized incidents is 1 mrem. Do these incidents really merit all this publicity? How dangerous is 1 mrem of radiation?

Perhaps the best way to understand this is to compare it with natural radiation—the 15,000 particles from natural sources that strike us each second throughout life. We are constantly bombarded from above by cosmic rays showering down on us from outer space giving us 30 mrem per year; from below by radioactive materials like uranium, potassium, and thorium in the ground—20 mrem/year; from all sides by radiation from the walls of our buildings (brick, stone, and plaster are derived from the ground)—10 mrem/year; and from within due to the radioactivity in our bodies (mostly potassium)—25 mrem/year. All of these combined give us a total average dose of about 85 mrem per year from natural sources,[3] or 1 mrem every 4 days. Thus radiation exposures in these highly publicized incidents are no more than what the average American receives every few days from natural sources.

But, you might say, the *extra* radiation is what we should worry about because there is nothing we can do about natural radiation. Not true. The numbers given above are national averages, but there are wide variations.[3] In Colorado and other Rocky Mountain states (Wyoming, New Mexico, Utah), where the uranium content in soil is abnormally high and where the high altitude reduces the amount of air above that shields people from cosmic rays, natural radiation is nearly twice the national average; but in Florida where the altitude is minimal and the soil is deficient in radioactive materials, natural radiation is 15% below the national average. Thus the radiation exposures in the highly publicized incidents are about equal to the extra radiation you get from spending five days in Colorado. Of course, millions of people spend their whole lives in Colorado, and it turns out that the cancer rate[9] in that state is 35% *below* the national average. Leukemia, probably the most radiation-specific type of cancer, occurs at only 86% of the National average rate in Colorado, and at 61% of that rate in the other high-radiation Rocky Mountain states. This is a clear demonstration that radiation is *not* one of the important causes of cancer. (Recall our estimate that it is responsible for only 0.5% of all fatal cancers.*)

Diagnostic X-rays are our second largest source of radiation exposure. A chest or a dental X-ray gives us about 10 mrem, but nearly all other X-rays give far higher exposures[10]: pelvis, 90 mrem; abdomen, 150 mrem; spine, 400 mrem; barium enema, 800 mrem, etc. Often a series of X-rays is taken, giving total exposures of several thousand millirems. The average American gets about 80 mrem per year[11] from this source, 80 times the exposure in the highly publicized radiation incidents. Again this diagnostic X-ray exposure is not unavoidable—much could be done to reduce it substantially without compromising medical effectiveness, and large numbers of X-rays are taken only to protect doctors and hospitals against liability suits.

There are several trivial sources of radiation that give us[12] about 1 mrem: an average year of TV viewing, from the X-rays emitted by television picture tubes; a year of wearing a luminous dial watch, since the luminosity comes from radioactive materials; and a coast-to-coast airline flight, because the high altitude increases exposure to cosmic rays. Each of these activities involves about the same radiation exposure as the highly publicized incidents.

How dangerous is 1 mrem of radiation? The answer can be given in quantitative terms: for each millirem of radiation we receive, our risk of dying of cancer is increased by one chance in 8 million.* This is the result arrived at independently by the U.S. National Academy of Sciences Committee on Biological Effects of Ionizing Radiation (BEIR)[13] and the United Nations Scientific Committee on Effects of Atomic Radiation (UNSCEAR).[11] The International Commission on Radiological Protection (ICRP) gives the risk as one chance in 10 million,[14] which is substantially in agreement.

This risk corresponds to a reduction in our life expectancy by 1.2 minutes.[15] A similar reduction in our life expectancy is caused by[15]

- crossing streets three times (based on the average probability of being killed while crossing a street)

---

* This estimate does not include the effects of radon, to be discussed later.
* This will frequently be referred to as the "cancer risk," which should be inferred as referring only to fatal cancers. Unless otherwise specified, a dose of radiation refers to exposure of the whole body at that level.

- taking about three puffs on a cigarette (each cigarette smoked reduces life expectancy by 10 minutes)
- an overweight person eating 10 extra calories (e.g., a small bite from a piece of bread and butter)
- driving an extra three miles in an automobile

These examples should put the risk of 1 mrem of radiation into proper perspective. Many more examples will be given in Chapter 4.

There has been intermittent publicity over the years for the fact that nuclear power plants, as a result of minor malfunctions or even in routine operation, occasionally release small amounts of radioactivity into the environment. As a result, people living very close to a plant receive about 1 mrem per year of extra radiation exposure. From the above example we see that, if moving away increases their commuting automobile travel by more than 3 miles per year (15 yards per day), or requires that they cross a street more than one extra time every four months, it is safer to live next to the nuclear plant, at least from the standpoint of routine radiation exposure.

SCIENTIFIC BASIS FOR RISK ESTIMATES

How do we arrive at the estimated cancer risk of low-level radiation, one chance in 8 million per millirem? We know a great deal about the cancer risk of *high*-level radiation, above 100,000 mrem, from various situations in which people were exposed to it and abnormally high cancer rates resulted.[13] The best known example is the carefully followed group of 80,000 Japanese A-bomb survivors, among whom 8500 people were exposed to doses in the range 100,000–600,000 mrem and suffered about 200 excess* cancer deaths up to 1974. In the period 1935–1954, it was fashionable in British medical circles to treat an arthritis of the spine called "ankylosing spondylitis" with heavy doses of X-rays, averaging about 300,000 mrem, and up to 1970 there were about 80 excess cancers among 14,000 patients so treated. In Germany that disease and spinal tuberculosis were treated with radium injections which administered 900,000 mrem to the bone; in a study of 900 patients treated before 1952, there were 54 excess bone cancers up to 1978. In Germany, Denmark, and Portugal, thorium (a naturally radioactive element) was injected

---

* In any study, there is a number of cases *observed* and a number of cases *expected;* the *excess* is the amount by which the observed exceed the expected. The "expected" number is the number normally found in a group of people similar in age, sex, race, and living conditions who were not exposed to the radiation in question; it is frequently obtained from national statistics.

into patients to aid in certain types of X-ray diagnosis between 1928 and 1955, giving several million millirem to the liver; among 3000 patients there were about 300 excess liver cancers up to 1977. Between 1915 and 1935, numerals on luminous watch dials were hand-painted in a New Jersey plant using a radium paint, and the tip of the brush was formed into a point with the tongue, thereby getting radium into the body; among 775 American women so employed, the average bone dose was 1.7 million millirem, and there were 48 excess bone cancers. Among 4100 American miners who were exposed to radon gas (a naturally radioactive material) in poorly ventilated mines between 1920 and 1968 and thereby received an average of 6 million millirem to their bronchial surfaces, there were 134 excess lung cancers up to 1974. There are also data from radon exposure to other groups of miners, British women given X-ray treatments for a gynecologic malady, women in a Nova Scotia tuberculosis sanatorium who were repeatedly subjected to X-ray fluoroscopic examination, American women given localized X-ray treatment for inflammation of the breasts, American infants treated with X-rays for enlargement of the thymus gland and other problems, Israeli infants treated with X-rays for ringworm of the scalp, natives of the Marshall Islands exposed to fallout from a nuclear bomb test, and many other groups.

From all of these data, there is good information on effects of high-level radiation (above 100,000 mrem). The problem is to use it to determine effects of low levels (below 10,000 mrem). The simplest idea is to assume that the risk is proportional to the dose—that the risk from 1 mrem is just 1/100,000 times the known risk from 100,000 mrem, the risk from 10 mrem is ten times larger than this risk from 1 mrem, and so forth. This is called the *linear hypothesis,* and with relatively slight modification, it is the method used to estimate effects of low-level radiation. The justification for the procedures used, some of the criticisms of them that have been highly publicized in the media, and responses to the latter are treated in the Appendix. The risk estimates obtained by these methods are almost universally accepted and used in the scientific literature, in government documents from all nations, in licensing of nuclear facilities, and in all other circumstances where health effects of low-level radiation are relevant.

THE MEDIA AND RADIATION

We now turn to the question of *why* the public is so fearful of radiation. Probably the most important reason is the gross overcoverage of radiation stories by the media. Even if the reporting is accurate, constantly hearing stories about radiation as a hazard gives people the subconscious impression that it is something to worry about. In attempting to document this overcoverage, I obtained the number of entries in the New York Times Information Bank on various types of accidents* and compared them with the number of fatalities per year caused by these accidents in the United States. I did this for the years 1974–1978 so as not to include the Three Mile Island accident, which generated more stories than usual. On an average, there were 120 entries per year on motor vehicle accidents, which kill 50,000; 50 entries per year on industrial accidents, which kill 12,000; and 20 entries per year on asphyxiation accidents, which kill 4500; note that for these the number of entries, which represents roughly the amount of newspaper coverage, is approximately proportional to the death toll they cause. But for accidents involving radiation, there were something like 200 entries per year, in spite of the fact that there has not been a single fatality from a radiation accident for the past 15 years.

From all of the hundred or so highly publicized incidents discussed earlier in this chapter (with the exception of the Three Mile Island accident) the total radiation combined received by all those involved was not more than 10,000 mrem[16]; since we expect only one cancer death from every 8 million mrem, there is much less than a 1% chance that there will ever be even a single fatality from all of those incidents taken together. (In the Three Mile Island accident, the sum of the doses to all members of the public[5] was 2 million millirem, so there is one chance in four that there will some day be a single fatality resulting from that accident.) On an average, each of these highly publicized incidents involves less than one chance in 10,000 of a single fatality, but for some reason they get the headlines instead of other accidents that kill an average of 300 Americans *every day,* seriously injuring ten times that number. Surely, then, the amount of coverage of radiation incidents has been grossly out of proportion to the true hazard.

Another problem in media coverage is use of inflammatory language. We often read and hear about "deadly radiation" or "lethal radioactivity," referring to a hazard that hasn't claimed a single victim for over 15 years, and has caused less than ten deaths in American history.[17] But we never hear about "lethal electricity," although 1200 Americans die each year from electrocution, or about "lethal natural gas," which kills 500 annually with asphyxiation.

A more important problem with media stories about radiation is that they never quantify the risk. I can understand their not giving doses in millirem—that may be too technical for their audience—but they could easily compare

---

* The numbers of entries were determined by the library of Argonne National Laboratory.

exposures with natural radiation or medical X-rays. In the 1982 accident at the Rochester power plant which was headline news for two days, wouldn't it have been useful to tell the public that no one received as much exposure from that accident as he receives every day from natural sources? This is not a new suggestion; similar comparisons have constantly been made by scientists for 35 years in information booklets, magazine articles, and interviews, but the media hardly ever use them. In the Rochester accident, they did everything to enhance the impression of danger, as in stating the number of people living in that area with the clear inference that that was the number of people being exposed to radiation. Occasionally, they quote a utility spokesman or a government bureaucrat saying that there is no danger to the public. They obviously attempt to give the impression that these people would have something to lose by telling the truth. Moreover, they always leave the audience with the impression that substantial harm to human health has occurred, although that has never been the case.

Another reason for public misunderstanding of radiation is that the media portray it as something very new and highly mysterious. There is, of course, nothing new about radiation because natural radioactivity has always been present on Earth, showering man with hundreds of times more radiation than he can ever expect to get from the nuclear power industry. The "mystery" label is equally unwarranted. Radiation effects are much better understood by scientists than those of air pollution, food additives, chemical pollutants in water, or just about any other agent of environmental concern. There are several reasons for this. Radiation is basically a much simpler phenomenon, with simple and well-understood mechanisms for interacting with matter, whereas air pollution and the others may have dozens or even hundreds of important components interacting in complex and poorly understood ways. Radiation is easy to measure and quantify, with relatively cheap and reliable instruments providing highly sensitive and accurate data, whereas instruments for measuring other environmental agents are generally expensive, erratic in behavior, insensitive, and inaccurate. And finally our knowledge of radiation health effects benefits from a $2 billion research effort under the nuclear power program, whereas only a tiny fraction of that amount has been devoted to health impacts of other pollutants. More important than the total amount of money is the fact that research funding for radiation health effects has been fairly stable for 40 years, thereby attracting good scientists to the field, allowing several successive generations of graduate students to be trained and excellent laboratory facilities to be developed.

As part of its effort to make radiation seem mysterious, the media give wide publicity to any scientist or any scientific work purporting to have evidence that makes it seem more dangerous. The Mancuso study[18] discussed in the Appendix is the best-known example of such work. In spite of the fact that it has been almost universally rejected by the scientific community (cf. Appendix), it is frequently referred to in the media. The New York Times Information Bank contains eleven entries on the Mancuso work,* but only a single entry on the 20-plus critiques[19-23] of it that have appeared in the scientific literature.

That one entry was a report of a meeting in Houston where Mancuso and one of his critics spoke. Mancuso's reply to the critic was that he had just completed a new analysis of the data which removed the critic's principal objection (and incidently eliminated most of the effect he had originally reported—the press report didn't mention that point). Of course, the critic had not seen the new analysis, for this was the first time it had ever been mentioned. He therefore could not criticize it. The press report was quite favorable to Mancuso, giving the impression that he had answered the critic and had done a new analysis comfirming the effect. When the new Mancuso analysis became available it still had plenty to criticize. This was forcefully pointed out in a study sponsored by the U.S. General Accounting Office[21] and carried out largely by the Statistics Department of the Massachusetts Institute of Technology. All official Committees of eminent scientists have either explicitly rejected or ignored the Mancuso study. I know of no radiation scientist not a party to the study who gives any credence to its results.† In a poll of radiation health scientists, Mancuso's credibility rating was particularly low (cf. Chapter 10). Yet many people whose information on radiation comes from the news media express surprise when I tell them that the Mancuso work is not accepted by the scientific community.

I remember the meeting of the Health Physics Society in Minneapolis in 1978 where Mancuso and some of his critics were scheduled to speak. The TV cameras were set up well ahead of time, and they operated continuously as Mancuso presented his arguments. But when he finished and his critics began to speak, the TV equipment was disassembled and carried away. This has been the rule rather than the exception, as anyone who attends these meetings can testify. I have been told that similar things happen routinely at Congressional hearings on radiation issues.

And this one-sided coverage goes far beyond scientific meeting rooms. When a dozen or so antinuclear protesters announce that they are going to

---

* These data on the New York Times Information Bank were obtained by the University of Pittsburgh library.

† One possible exception is K. Z. Morgan, whom I exclude because he had some involvement in the Mancuso Study on a consulting basis.

stage a protest outside a nuclear plant, the TV cameras dutifully appear to record their attempting to enter the plant and being arrested. A nationwide audience views the scene. By contrast, I know of several pro-nuclear rallies attended by many hundreds of people—one in Colorado drew 5000—with U.S. Senators and similar nationally prominent personalities as speakers, but I have never seen national TV or magazine coverage of them. Since only this type of coverage can have an impact on public opinion, biased coverage is a powerful force encouraging antinuclear protests. Once again, a distorted picture is projected to the public.

To return to the Mancuso study, it is important to recognize that the issue here is a *strictly scientific* one—what is the cancer risk of one millirem of radiation? It is most difficult to see how there can be political or other nonscientific aspects to that question. Nevertheless, the media consistently treat it as a political question.

There has been one other recent paper in the scientific literature claiming evidence that this risk is about ten times larger than the one chance in 8 million estimate by the official committees. That paper, by Bross and collaborators,[24] was immediately followed in the scientific journal by a devastating critique of it by two eminent scientists from the National Cancer Institute.[25] The Bross paper was given extensive coverage by the *New York Times* and other newspapers all over the country, but there was not a single word about the critique—that, apparently was not included among "all the news that's fit to print." Again the Bross paper was evaluated and explicitly rejected by the National Academy of Sciences BEIR Committee, and has been ignored by all other official committees. Again there has been no mention of these rejections by the media.

There are, of course, papers published in scientific journals presenting evidence that makes low-level radiation seem *less* dangerous* than the estimates by the official committees. Some even ascribe *beneficial* effects on human health* to the low levels generally encountered in connection with nuclear power.[26] But these reports never receive media coverage.

---

* These papers are good science; they collect and correctly analyze data. However, it should not be inferred that any one paper gives a definitive answer on the cancer risk of low-level radiation. For example, a study of the cancer risk for mice may not be applicable to humans. No study of humans can be carefully controlled; other factors, such as variations in diet, exposure to chemicals, smoking habits, socioeconomic status, etc. are frequent complications. Each study gives evidence, of variable relevance, and taken together they form a distribution of justifiable risk estimates. The scientific consensus would ordinarily be near the center of this distribution. However, in the cases of health risks, there is a long-standing tradition of choosing risk estimates nearer to the maximum that is scientifically justifiable, as it is felt that it is better to overestimate a health risk than to underestimate it.

Since Mancuso and I are at the same university, I tried a test of "the system." Shortly after his paper had drawn the tremendous publicity described above, I happened to be publishing a paper in the same journal[27] which gave evidence that radiation is somewhat less dangerous than the accepted estimates. I wrote a highly enthusiastic press release on my paper and arranged to have it put out by our University Public Relations Office, the same one that had handled Mancuso's paper, but I never got a single call from the media. The only difference media people could possibly have been able to distinguish between Mancuso's paper and mine from the press releases was that his said that radiation was more dangerous, while mine said it was less dangerous.

One of the most disturbing aspects of the media's handling of nuclear energy questions is that they are effectively making decisions on strictly scientific questions. To see how this happens, consider a hypothetical situation in which the color of an object, ranging from black through various shades of gray to white, is an important question of public concern that can only be determined through complex scientific investigation. Let us say that a thousand scientists spend several years studying the question, publishing papers on it, and discussing it at meetings. Eventually 10 believe that the object is black, 20 believe it is some dark shade of gray, 70 believe it is some light shade of gray, and the remaining 900 believe it is white. Surely the consensus of the involved scientific community is that it is white or close to it. But suppose a TV producer, because of political prejudices, believes the object is black and decides to do a special on the subject. He interviews five of the ten scientists who say that the object is black and five government bureaucrats who, having administered the research, say that its conclusion was that the object is white. These interviews often last for two hours or more, from which the producer selects perhaps two minutes from each interview and juxtaposes these segments to suit his purposes. Government bureaucrats have low credibility to many people. The producer can easily select segments where they have facial expressions that make them *seem* insensitive, stupid, or bureaucratic, or which, in any of a dozen other ways, turns people off. For the five scientists advocating the color black, the producers can choose segments making them seem intelligent, human, and very scientific—image building or breaking is part of the stock-in-trade of TV producers. At the end of the program there is a summary by the TV host, saying that the scientific community is split on the issue, so the public must judge which is correct. Note that there was never an indication that the split was 900 versus 10, and from all evidence given on the program one would get the impression that it is something like 50–50. The government bureaucrats may have said many times that it was 900–10, but those segments never got on the air.

The TV producer says, correctly, that he gave equal time to both sides of the issue. He claims that he was only presenting the information, leaving the public to decide. But how would the public be expected to react? Most people would curse the damned government bureaucrats—"those bastards are trying to deceive us"—and bless the courageous scientists who stood up against them. "Of course the object is *black*." Thus, on a strictly scientific question, the scientific community reaches a consensus that the object is white, while the public becomes convinced that it is black. The TV producer, rather than the scientific community, has decided upon the answer the public accepts.

This story may seem highly hypothetical and exaggerated, but it is exactly what has happened on the issue of the dangers of radiation. The "white color" corresponds to the position of the various official committees of prestigious scientists like BEIR, UNSCEAR, ICRP, and NCRP, while the "black color" corresponds to the positions of a very few antinuclear scientists like Mancuso, Sternglass, and Gofman. The numbers on each side are quite realistic, as shown in Chapter 10. The TV special corresponds to any one of a dozen that have been aired over the past few years, or to many segments of programs like "60 Minutes," "20–20," network news, or TV magazine programs.

Thus, it has come to pass that the public's estimate of the cancer risk of 1 mrem of radiation, a strictly scientific question if there ever was one, is not that of the National Academy of Sciences Committee, not that of the United Nations Scientific Committee, and not that of the International Commission on Radiological Protection, all three of which agree, but rather that of TV reporters with no scientific education or experience, often heavily influenced by political prejudices regarding nuclear power. I'll never forget the reporter who told me that he didn't care about National Academy Committees or such—he had spoken to Mancuso and could tell that what he was saying was right.

It is my impression that media people consider the official committees of scientific experts to be tools of the nuclear industry. It is difficult for me to understand that attitude. The National Academy of Sciences is a nonprofit organization chartered by the U.S. Congress in 1863 to further knowledge and advise the Government. It is composed of about a thousand of our nation's most distinguished researchers from all branches of science. It appoints the Committee on Biological Effects of Ionizing Radiation (BEIR) and reviews its work. The BEIR Committee itself is composed of about 21 American scientists well recognized in the scientific community as experts in radiation biology; 13 of them are university professors, with lifelong job security guaranteed by academic tenure. The United Nations Scientific Committee on Effects of Atomic Radiation (UNSCEAR) is made up of scientists from 20 nations from both sides of the iron curtain and the Third World. The nations with current representation on the International Commission on Radiological Protection (ICRP) are similarly distributed, and the current chairman is from Sweden. It is extremely unusual for such highly reputable scientists to practice deceit, if for no other reason than that it would be easy to prove that they had done so and the consequences to their scientific careers would be devastating. All of them have such reputations that they could easily obtain a variety of excellent and well-paying academic positions independent of government or industry financing, so they are not vulnerable to economic pressures. But above all, they are human beings who have chosen careers in a field dedicated to protection of the health of their fellow human beings; in fact many of them are M.D.s who have foregone financially lucrative careers in medical practice to become research scientists.* To believe that nearly all of these scientists are somehow involved in a sinister plot to deceive the public is indeed a challenge to the imagination.

The media have frequently stated that the scientific estimates of the cancer risk of low-level radiation have been increasing in recent years. This is definitely not the case. The BEIR estimate was reduced from one chance in 6 million per millirem[28] in 1972 to one chance in 8 million in 1980,[13] and the UNSCEAR and ICRP estimates of cancer risk from radiation have not changed in the past decade.

It is only natural to wonder *why* the media so heavily overcover and distort information about radiation. One possible reason is that they are in the entertainment business and educating the public is not their primary concern. I have been told that one point in the Nielsen rating for the TV network evening news is worth $7 million per year in advertising revenue. In that atmosphere, what would happen to a TV producer who decided to concentrate on properly educating the public in preference to entertaining it? Somehow the media have found that scaring the public about radiation attracts viewers, so they play that game to the hilt. As an illustration of the low priority they place on their educational function, I doubt if there are more than one or two Ph.D.-level scientists in the full-time employ of any television network, in spite of the fact that they are the primary source of science education for the public. Even a strictly liberal arts college with no interest in training scientists would have one Ph.D.-level scientist for every 200 students, whereas the networks have practically none for their 200 million students.

---

* Radiation health scientists rarely have opportunities to consult for industry; if they were out to make more money, they would do very well to join the antinuclear forces, who are well financed and pay generous lecture fees.

The situation would be tolerable if the media considered it their function merely to transmit scientific information from the scientific community to the public. But this they refuse to do. They want to decide what to transmit, which requires media people to make judgments on scientific issues. How they do this clearly depends heavily on their own education on these issues, which is derived almost exclusively from other media sources. Thus the media fraternity dealing with science seems to form a closed loop, educating one another, and choosing input from scientists only when it suits their own prejudices.

When I complain about this to media people, they always say that the scientific community is split on the issue. By "split" they seem to mean that there is at least one scientist disagreeing with the others. They don't seem to recognize that a unanimous conclusion of a National Academy of Sciences Committee should be given more weight than the opinion of an individual scientist. Yet if that individual has a viewpoint that suits their prejudices it gets all their coverage. It reminds one of the Russian government deciding that the genetic theories of Lysenko were correct because it suited the Communist Party philosophy—never mind that essentially all scientific geneticists rejected it.

The total number of scientists to have publicly claimed that the risks given by the official committees are gross underestimates is less than 10, representing less than 1% of the involved scientific community. But this tiny unrepresentative group has received the lion's share of the media scientific coverage on effects of radiation. Moreover, there is disagreement within this tiny minority. Some make claims only about leukemia while some claim that certain other forms of cancer are extremely sensitive to radiation but not leukemia. None of them publishes much in scientific journals, and they seldom refer to one another's work.

Some media people complain that, since the scientific community is split, they have no way to find out what the scientific consensus is. To this I always propose a simple solution: pick a few major universities at random, call and ask the operator for the department chairman or a professor in the field, and ask the question; after five such calls the consensus will be clear on most questions, usually 5 to 0. The media people never seem willing to do this. My strong impression is that they aren't interested in what scientists have concluded, but only want to support their preconceived viewpoints.

The poll of radiation health scientists discussed in Chapter 10 show clearly that the scientific community is *not* split. It has a very strong consensus. The idea of a split is largely a creation of the media, and they use it to avoid telling the public about that consensus.

GENETIC EFFECTS OF RADIATION[29]

Other than inducing cancer in rare instances, the only significant health impact of low-level radiation is causing inherited disabilities in later generations. These disabilities, often called *genetic defects,* range from minor problems like color blindness to very serious maladies like mongolism. There is further discussion of their nature in the Appendix. The news media have not emphasized this aspect, and the public's fear of it is seldom spoken, but in personal conversations, I have often derived the impression that it is deeply felt and dreaded. Some people believe that radiation can produce two-headed children, or various types of subhuman or superhuman monsters. This is obviously not the case because they have never been produced before in spite of the fact that humanity has always been exposed to natural radiation, not to mention the few generations that have been exposed to medical X-rays, sometimes in very heavy doses (millions of millirem).

Some people have the impression that radiation-induced genetic effects can destroy the human race, but that is also false. The law of natural selection causes good traits* to be bred in and bad traits to be bred out, so any bad traits induced by radiation will eventually disappear. The genetic effects of the nuclear industry can best be understood by recognizing that it increases man's radiation exposure by less than 1% over that from natural radiation, and natural radiation is believed to be responsible for only about 3% of all normally encountered genetic defects.[13] Thus the impact of a large-scale nuclear industry would increase the frequency of genetic defects by less than 1% of 3%, or one part in 3000.

Probably an easier way to understand the genetic effects of nuclear power is to make a comparison with other human activities that induce genetic defects. One good example is parents having children at older ages, which increases the likelihood of several types of genetic disease. Increased maternal age is known[30] to enhance the risk of Down's syndrome, Turner's syndrome, and several other chromosomal disorders, while increased paternal age rapidly raises the risk of achondroplasia and presumably a thousand other autosomal dominant diseases.[31] The genetic effects of a large nuclear industry would be equal to those of delaying the conception of children by an average of 2.6 days.[29] Between 1960 and 1973, the average age of parenthood increased by about 50 days, causing 20 times as much genetic disease as would be induced by a large nuclear industry.

To be quantitative about genetic effects of radiation, we can expect one

---

* "Good traits" here means inherited traits that help the human species to survive.

genetic defect in all future generations combined for every 11 million millirem of individual radiation exposure to the general population.[13] For example, natural radiation exposes the average American to 85 mrem per year, or a total of (85 × 230 million =) 20 billion millirem per year to the whole population. It can therefore be expected to cause (20 billion ÷ 11 million =) 2000 cases per year of genetic disease, about 2% of all the cases normally occurring. (Since our present population includes an abnormally large number of younger people, this result is increased to 3%.) This is the estimate of the National Academy of Sciences BEIR Committee, and those of UNSCEAR[10] and ICRP[14] are similar. It is interesting to point out that these estimates are derived from studies on mice, because there is no actual evidence for radiation causing genetic disease in humans. The best possibility for finding such evidence is among the survivors of the A-bomb attacks on Japan, but several careful studies have found no evidence for an excess of genetic defects among the first generation of children born to them.[13] If humans were appreciably more susceptible to genetic disease than mice, a clear excess would have been found.* We can therefore be confident that in utilizing data on mice to estimate effects on humans, we are not understating the risk.

Often an individual worries about his own personal risk of having a genetically defective child; it is about one chance in 40 million for each millirem of exposure receiving prior to conception—somewhat more for men and somewhat less for women. This is equal to the risk of delaying conception by 1.2 hours.[29]

It may be relevant here to mention that air pollution and a number of chemicals can also cause genetic defects. There is at least some mutagenic information on over 3500 chemicals[33] including[34] bisulfites, which are formed when sulfur dioxide is dissolved in water (these have caused genetic changes in viruses, bacteria, and plants), and nitrosamines and nitrous acid, which can be formed from nitrogen oxides. Sulfur dioxide and nitrogen oxides are the two most important components of air pollution from coal burning. Other residues of coal burning known to cause genetic transformations are benzo-a-pyrene (evidence in viruses, fruit flies, and mice), ozone, and large families of compounds similar to these two. The genetic effects of chemicals on humans are not well understood and there is practically no quantitative information on them, but there is no reason to believe that the genetic impacts of air pollution from coal burning are less harmful than those of nuclear power.

---

* There is one paper[32] which, based on data for the Japanese A-bomb survivors, concludes that humans are 4 times less susceptible to genetic damage than mice. However, the geneticists I consulted did not judge its evidence to be firm enough for widespread adoption.

Caffeine and alcohol are known to cause genetic defects. One study[35] concludes that drinking one ounce of alcohol is genetically equivalent to 140 mrem of radiation exposure, and a cup of coffee is equivalent to 2.4 mrem.

While the media have generally played up the fear of cancer more than genetic effects in scaring the public about radiation, they have not entirely ignored the latter. There was one TV special which featured two beautiful twin babies (dressed in very cute dresses) afflicted with Hurler's syndrome, a devastating genetic disease. All sorts of details were offered on its horrors— they will go blind and deaf by the time they are five years old, and then suffer from problems with their hearts, lungs, livers, and kidneys, before they die at about the age of ten. Their father, who had worked with radiation for a short time, told the audience that he was sure that his radiation exposure had caused the genetic disease of his children. What better tear-jerking propaganda could there be to convince the audience that radiation is a horrible thing? There was no mention of the fact that the father's total occupational radiation exposure was only 1300 mrem, less than half of his exposure to natural radiation up to the time his children were conceived. With that much exposure, the risk of a child deriving a genetic defect is one chance in 25,000; their normal risk is 3%, due to spontaneous mutations, so there is only once chance in a thousand that their genetic problems were due to their father's job-related radiation exposure.

Incidentally, the *New York Times Magazine* ran a story featuring these same babies, complete with pictures, and giving the reader every reason to believe that their genetic problem resulted from their father's radiation exposure. Again, the fact that there is only one chance in a thousand that this is the case was apparently not considered to be "news that's fit to print."

Perhaps the most important human activity that causes genetic defects is the custom of men wearing pants.[36] This warms the sex cells and thereby increases the probability for spontaneous mutations,* the principal source of genetic disease. Present very crude estimates are that the genetic effects of 1 mrem of radiation are equivalent to those of 5 hours of wearing pants.[29]

If one is especially concerned with genetic effects, there is much that can be done to reduce them. By using technology now available like amniocentesis, sonography, and alpha fetoprotein quantification, we could avert 6000 cases per year of genetic defects in the United States, at a total cost of

---

* Spontaneous mutations are basically chemical reactions. The fact that chemical reactions are speeded up by heat is familiar from cooking, washing with hot water, use of heat pads to promote healing, using a match to start a fire, etc.

$160 million in medical services.[37] By comparison, a large nuclear power industry in the United States would eventually cause 37 cases per year while producing a product worth $50 billion and paying about $8 billion in taxes. Thus, if 2% of this tax revenue—$160 million—were used to avert genetic disease, it could be said that the nuclear industry is averting $(6000 \div 37 =)$ 160 cases for every case it causes.[29] If the money were spent on genetic research, it would be even more effective.

One bothersome aspect of the genetic impacts of nuclear power is the conscience-burdening idea that we will be enjoying the benefits of the energy produced while future generations will be bearing the costs. However, we must recognize that there are many other, and much more important situations in which our generation and its technology are adversely affecting the future. Perhaps the most important is our consumption of oil, gas, coal, and other mineral resources mentioned in Chapter 1; we are also overpopulating the world, exhausting agricultural land, developing destructive weapons systems, and cutting down forests. By comparison with any one of these activities, the genetic effects of radiation from the nuclear industry are exceedingly trivial. Moreover, this is not a new situation: forests in many parts of Europe were cut down at a time when wood seemed to be the most important resource for both energy and structural materials, and local exhaustion of agricultural land, game, and fish stocks has been going on for millenia.

Nevertheless, at least in recent times, each succeeding generation has lived longer, healthier, and more rewarding lives. The reason, of course, is that each new generation receives from its predecessors not only a legacy of detriments, but also a legacy of benefits. We leave to our progeny a tremendous fund of knowledge and understanding, material assets including roads, bridges, buildings, transport systems, and industrial facilities, well organized and generally well functioning political, economic, social, and educational institutions, and so on, all far surpassing what we received from our forbears. The important thing from an ethical standpoint is *not* that we leave our progeny *no* detrimental legacies—that would be completely unrealistic and counterproductive to all concerned—but rather that we leave them more beneficial than detrimental legacies.

In the case of the nuclear industry and the genetics effects it imposes on later generations, any meaningful evaluation must balance the value to future generations of an everlasting source of cheap and abundant energy developed at a cost of tens of billions of dollars and tens of thousands of man-years of effort, against a few cases of genetic disease which we also leave them the tools to combat cheaply and efficiently.

OTHER HEALTH EFFECTS OF RADIATION

When lecturing about radiation effects, I am often asked whether there aren't other than cancer and genetic defects induced. Careful studies[38] among the survivors of the atomic bomb attacks on Japan, where 24,000 people were exposed to an average of 130,000 mrem each, revealed no such evidence. Moreover, our understanding of how various diseases develop leads us not to expect other diseases from low-level radiation.

One other effect of relatively low levels of radiation is developmental abnormalities among children exposed to radiation *in utero*.* This is well known from animal studies, and there is also extensive human evidence from medical exposures and from studies of the Japanese A-bomb survivors.[39] Among the latter, children exposed prior to birth to more than 25,000 mrem were, at age 17, an average of 0.9 in. shorter, 7 lb lighter, and nearly a half inch smaller in head diameter than average. Among the 22 children who received more than 150,000 mrem from the bomb before the eighteenth week of gestation, 13 have small head size and 8 have mental retardation.[40] There were only two cases of mental retardation among those exposed after the eighteenth week. There was no evidence for mental retardation among those exposed to less than 50,000 mrem.[41] From animal studies it is expected that there may be slight developmental abnormalities for doses as low as a few thousand millirem at critical times during fetal development. (A woman living very close to a nuclear power plant would typically receive only about 1 mrem during pregnancy.)

There has also been extensive consideration of the possibility that *in utero* exposure may give a large risk of childhood cancer.[39] It was initially reported that children whose mothers received pelvic X-rays during pregnancy had tenfold elevated risks for this disease.[42,43] However, many factors other than X-rays have been found to be similarly correlated with childhood cancer, including[39] use by the mother of aspirin and of cold tablets, and the child's blood type, viral infections, allergies, and appetite for fish and chips (this was a British study). It was shown[44] that the original correlations would have predicted 18 excess childhood cancers among those exposed *in utero* to the A-bomb attacks on Japan, whereas none occurred. The initial observations could be explained as effects of the medical problems that required the mother's X-rays. A study[45] of mothers given X-rays for nonmedical reasons showed no evidence for increased cancer. Nevertheless, the official committees con-

---

* Genetic effects, discussed previously, are due to damage to sex cells of parents prior to conception. The effects discussed here are due to damage to the fetus long after conception.

sider this excess cancer risk to be a possibility; it is therefore taken into account in setting radiation protection regulations for occupational exposure of pregnant (or potentially pregnant) women. From the public health viewpoint, *in utero* exposure is not of great importance since we spend only 1% of our lives *in utero*.

These *in utero* effects are often exaggerated in the public mind. There were widely circulated pictures of grossly deformed farm animals with claims that they were caused by *in utero* exposure to radiation from the Three Mile Island accident. There were, of course, no such claims in the scientific literature. Recall that average exposures from that accident were 1.2 mrem, which is equal to the radiation received from natural sources every five days, so even if the deformities were due to radiation, they would much more likely be due to natural radiation. There have been large numbers of careful experiments on effects of radiation to various animals—I have visited farms operated for that purpose in Tennessee and in Idaho—and no such effects were observed in those experiments from exposures less than several thousand millirem. It is therefore extremely hard to believe that such effects can occur at 1.2 mrem, or even at the maximum exposures very close to the Three Mile Island plant which were still well below 100 mrem. Naturally, a small fraction of all animals (or humans) born anywhere and under any circumstances are deformed, so there would be no great difficulty in collecting pictures of deformed animals. Before one can claim evidence that these effects are due to radiation from the Three Mile Island accident, however, it would have to be shown that the number of deformed animals in that area at that time was larger than the number in that area at previous times, or in other areas at that time, by a statistically significant amount. That has certainly never been shown for any health effects among humans or animals.

PUBLIC INSANITY

Because of the factors we have been discussing and perhaps some others, the public has been driven *insane* over fear of radiation. I use the word "insane" purposefully since one of its definitions is loss of contact with reality. The public's understanding of radiation dangers has virtually lost all contact with the actual dangers as understood by scientists. Perhaps the best example of this was the howl of public protest when plans were announced more than a year after the accident at Three Mile Island to release the radioactive gas that had been sealed inside the containment structure of the damaged reactor. This was important so that some of the safety systems could be serviced, and it was obviously necessary before recovery work could begin. Releasing this gas would expose no one to as much as 1 mrem, and the exposure to most of the protesters would be a hundred times less. Simply traveling to a protest meeting exposed the attendees to far more danger than release of the gas; moreover, an appreciable number fled the area, traveling a hundred miles or more, at the time of the release. Recall that 1 mrem of radiation gives the same risk as driving 3 miles or crossing a street 3 times on foot. Needless to say, the statements of fear by the protesters were transmitted to the national TV audience with no accompanying evidence that their fears were irrational.

One disheartening aspect of that episode was the Nuclear Regulatory Commission's efforts to handle it. An early survey of the local citizenry revealed that there was substantial fear of releasing the gas. The NRC therefore undertook a large program of public education, explaining how trivial the health risks were. When this public education campaign was completed, another poll of the local citizenry was taken. It showed that the public's fear was greater than it was before the campaign. The public's insanity on matters of radiation defies all rational explanation.

One tragic consequence of this public insanity is the impact it is having on medical uses of radiation. Radioactive materials and accelerator radiation sources are widely used for medical diagnosis and therapy, saving tens of thousands of lives each year. Recall from Chapter 1 that even the antinuclear activists claim an average of only 120 lives lost per year from all aspects of nuclear power whereas most estimates are more than ten times lower. Thus artificial radiation is saving hundreds or thousands of times as many lives as it is destroying. But as a result of the public insanity, patients are refusing radiation procedures with growing frequency, and physicians are becoming more hesitant to use them. In less than an hour of discussion during intermissions at a local meeting on nuclear medicine in New York City, I learned of the following situations:

- In one big-city hospital, about 20% of the patients refuse the recommended treatment for hyperthyroidism, which involves the use of radioactive iodine, opting instead for a less cost-effective drug treatment that frequently leads to relapse.[46]
- In another hospital nursery intensive care unit, when a portable X-ray machine is brought into the unit, the nurses leave the area, abandoning the infants under their care; it is estimated that remaining at their posts would expose them to less than 1 mrem per year of additional radiation.[47]
- A large medical center planned a project involving radiation for 1100 patients per year in its Intensive Care Unit. It was estimated that the

ten nurses working there would each be exposed to about 100 mrem per year of additional radiation, approximately the extra amount received by living in Colorado. The nurses threatened to strike, forcing abandonment of the project.[48]

I don't know of any elaborate quantitative estimates of the percentage by which fear is reducing medical use of radiation, but even if it is only 10%, this public insanity is killing at least many hundreds, and probably several thousand Americans each year.

APPENDIX

The Cancer Risk from Low-Level Radiation

As enumerated earlier in this chapter, there is a great deal of information on the cancer risk from high-level radiation, about 100,000 mrem. The problem is to use it to determine the risk at low levels, usually below 10,000 mrem. The simplest idea is to assume that the risk is proportional to the dose—that the risk from each millirem is just 1/100,000 times the known risk from 100,000 mrem. This is called the *linear hypothesis* and is represented by the straight line in Fig. 1. Two other possibilities are also shown in Fig. 1. The majority opinion in the scientific community is that the linear hypothesis is an upper limit on the effects of low doses that probably overestimates the risk; that is, the risk lies somewhere between the straight line and line A in Fig. 1 which curves upwards. There is a minority opinion, however, which favors the curve bending downward in Fig. 1, according to which the risk from low-level radiation is much greater. There is reason to believe that the

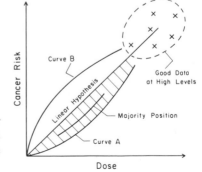

Fig. 1. Schematic representation of cancer risk versus dose. The good data are at large doses. Curves A and B are discussed in the text.

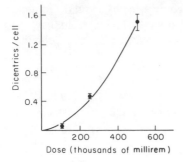

Fig. 2. Frequency of dicentrics (chromosome aberrations) in human white blood cells after exposure to gamma rays. From Ref. 52.

shape of the curve is different for various types of radiation, so we will confine most of our discussion here to the curve for gamma rays, beta rays, and X-rays, the most important type for most purposes.

The evidence supporting the majority position is as follows:[49,50]

1. There is a widely accepted theory of how radiation induces cancer[51] and a reasonable degree of experimental verification for many of its aspects. This theory predicts a line curving upward like line A in Fig. 1.

2. Studies of chromosome damage by radiation in human white blood cells indicate a behavior[52] like line A of Fig. 1. An example of data of this type[53] for cells grown in cultures is shown in Fig. 2.

3. Transformation to malignancy due to radiation in mouse embryo cells grown in culture generally follows a curve like line A of Fig. 1. Some examples[54,55] are shown in Fig. 3.

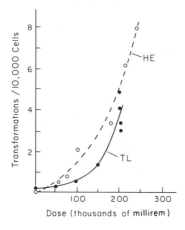

Fig. 3. Incidence of malignant transformation in C3H mouse embryo cell cultures after X-ray exposure. "TL" is from Ref. 53 and "HE" is from Ref. 54.

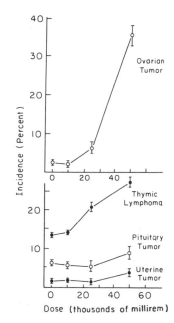

Fig. 4. Incidence of various cancers versus dose in gamma ray exposure of mice. From Ref. 55.

4. There are rather good data for cancer incidence in mice exposed to various doses of gamma rays, and the results generally indicate a behavior like line A of Fig. 1. Examples[56] are shown in Fig. 4.

5. The data on cancer incidence in mice injected with radioactive material show a behavior typified by the data[57] in Fig. 5. As with line A in Fig. 1,

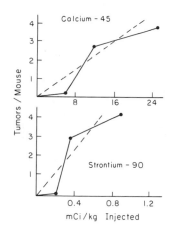

Fig. 5. Bone cancers produced in CF1 female mice by intravenous injection of strontium-90 and calcium-45, versus quantity injected in millicuries (mCi) per kilogram (kg) of mouse weight. From Ref. 56.

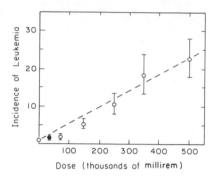

Fig. 6. Relative mortality rate from leukemia among Japanese A-bomb survivors. Plotted from data in Ref. 37.

use of a linear hypothesis based on the high-dose data, as shown by the dashed lines, grossly overestimates effects at low doses.

6. The best human data are for leukemia among the Japanese A-bomb survivors,[38] shown in Fig. 6 with a straight line to the high-dose data. The low-dose points are well below the line and there is only a few percent probability that this much deviation can be produced by statistical fluctuation.

7. Data on the radium watch dial painters,[58] shown in Fig. 7, indicate the same effect; again there is only a few percent probability that the deviations of the low-dose data from the straight line can be due to statistical fluctuations.

8. For lung cancers induced by radon,[59] a straight line to the high-dose data for miners predicts an incidence rate due to naturally occurring radon in the environment (to which everyone is exposed) that is larger than the actual lung cancer rate among non-cigarette-smokers. Moreover, the type of lung

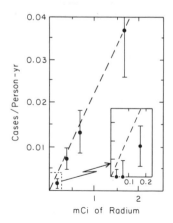

Fig. 7. Bone cancer incidence among radium dial painters. Main plot shows higher dose data used to obtain straight line, and inset compares this straight line with the low dose data. From Ref. 57.

cancer found predominantly among the miners is very rare among nonsmokers so only a small fraction of cases among nonsmokers can be due to radon. Thus the low-dose risk from radon is far below that given by the straight line, indicating a behavior like line A in Fig. 1.

9. There is abundant evidence that the time delay between exposure to radiation and the consequent development of cancer increases as the dose decreases. Evidence for this from animal experiments[60] is shown in Fig. 8. Dogs whose bones were exposed to 100 million millirem or more died of bone cancer within about 3 years, but when exposed to less than 5 million millirem they did not develop bone cancer until 9 years later. From these data it appears that with much lower exposure they would live their entire life span (11 years) and die from other causes before the cancer would develop. This effectively eliminates the risk of low-level radiation. There is evidence of this effect for the German ankylosing spondylitis patients treated with radium,[61] for skin cancer in Japanese radiological workers,[62] for the American radium-dial painters,[63] for leukemia in mice, for skin and bone cancer in mice and rats, and for breast cancer in rats.[64]

All of this evidence supports the majority position shown in Fig. 1, that the linear hypothesis does not underestimate, and very probably *over*estimates the risk of low-level radiation.

The most widely publicized evidence for the opposite position, represented by curve B, bending downward in Fig. 1, is based on a study by Mancuso and co-workers[18] of radiation workers in the U.S. Government's Hanford Laboratory in central Washington state. In comparing deaths among the workers with higher radiation exposures (typically 20,000 mrem) with those who had little or no exposure, they found that the former group died more frequently from cancer, especially of certain types, than the latter group. It should be noted that the Mancuso study only considered the workers who

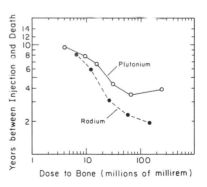

Fig. 8. Survival time for beagle dogs who developed bone cancer from injections of plutonium and radium versus dose received by their bone up to one year before death. From Ref. 59.

had died, paying no attention to the far greater number that were still alive. When all workers, living and dead, were considered, the most highly exposed group did *not* have a higher cancer risk than those with low exposure.[19] A large number of methodological shortcomings in the Mancuso work were pointed out. In all, that paper drew well over 20 critiques,[19-23] an unprecedented, large number. The original data on which their study was based were reanalysed independently by three other research teams.[19-21] All agreed that the only evidence from these data that might indicate effects from radiation were the following: among the most highly exposed workers, there were 3 cases of pancreatic cancer versus 1 expected, and there were 3 cases of multiple myeloma versus 0.4 expected.

Searches for evidence for increased incidence of these two diseases were made in other groups exposed to radiation, but no such evidence was found. For example, Fig. 9 shows[23] the data for pancreatic cancer among the Japanese A-bomb survivors along with the Mancuso analysis and the "3 observed versus 1 expected" found in the other analyses. It is evident that the A-bomb survivor data in the same dose region has much better statistical accuracy, showing no indication of the effect claimed by Mancuso.

Other evidence for radiation-induced multiple myeloma was even more negative. There were only three cases more than would be normally expected among the Japanese A-bomb survivors[65] whereas well over a hundred would have been expected from the Mancuso results. There were no cases among the German patients treated with radium for ankylosing spondylitis,[21] whereas the Mancuso conclusions would have predicted several hundred.

There is no difficulty in explaining the excess multiple myeloma and pancreatic cancer among the most exposed Hanford workers in other ways. For example, those exposed to higher radiation were technicians who are also

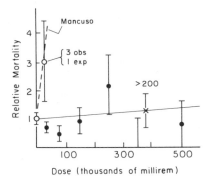

Fig. 9. Mortality from cancer of the pancreas among Japanese A-bomb survivors (black circles), among Hanford workers (open circles), and according to Mancuso analyses. Point labeled >200 represents the average of all data above 200,000 mrem. From Ref. 22.

exposed to various chemicals and unusual working environments, whereas the low-exposure group probably consisted mainly of office workers.

As a consequence of some of the problems we have pointed out and many others, the Mancuso study has been rejected[13] or ignored, after extensive evaluation, by all official scientific committees charged with responsibility in health effects of radiation. None of them use it in any way in discharging their responsibilities.

The second most highly publicized evidence purporting to indicate that radiation is extra dangerous came from studies of workers at the Portsmouth, New Hampshire naval shipyard which services nuclear submarines. Tom Najarian, a Boston physician, in talking to a patient, got the impression that there was an excess of leukemia among workers there. With the help of reporters from the *Boston Globe,* he searched through 90,000 death certificates. He concluded, in a report headlined in that newspaper,[66] that there were 22 deaths from leukemia among those workers whereas only 5 were expected.* In a publication in a scientific journal,[67] he later reported that there were 20 cases versus 11 expected—he never clearly explained how he determined the number expected and that number is very sensitive to the ages of the workers, which he did not know. From telephone conversations with surviving members of the families, he concluded that 6 of the leukemia victims had been exposed to radiation whereas only 1.1 case was expected among those so exposed; shipyard records later revealed that 3 of the 6 had no exposure. The National Center for Disease Control spent over a million dollars studying the question, and concluded that there is no evidence for health effects of radiation among the shipyard workers.[68] In a Congressional hearing chaired by Sen. Edward Kennedy, Najarian withdrew most of his claims, and the others were heavily critiqued by scientists from National Cancer Institute. Senator Kennedy, who is well known for his antinuclear sentiments, chastized Najarian for unduly alarming families of the shipyard workers.[69]

In an effort to see how the news media covered the Najarian story, I used the *New York Times Index* to find all stories on it. After Najarian's claims of large cancer excesses were published in the *Boston Globe,* he testified on them before Congressional committees, Congressmen and other politicians called for investigations, government agencies considered how the

---

* "Expected" normally means the number found in statistical data for people similar in age, sex, race, socioeconomic status, geographical habitat, etc. In many cases data for all Americans (or from the same state) of the same sex and age distribution are used. Age distribution is very important as leukemia incidence increases very rapidly with age, tenfold from age 35 to age 70.

investigation should be carried out and by whom, and there was other such activity over a one-year period. There were 14 articles in the *New York Times* over this period, most of them reiterating that there were a large number of excess cancers among the shipyard workers. There were then no further articles for nearly two years while the National Center for Disease Control was doing its study. When the study was concluded, reporting *no* evidence for excess cancers, there was a single story on page 37 of the Jan. 25, 1981 edition. The first 16 lines were introductory, reviewing the evidence that there were excess cancers; there were then 9 lines giving the results of the $1 million study which concluded that there were no such effects; and the story concluded with 15 lines casting doubt on the study. (In the scientific community, the study was accepted as definitive.) There was no report in the *New York Times* on the abovementioned hearings before Senator Kennedy's Committee. In summary, $14\frac{1}{2}$ of the 15 articles in the *New York Times* left readers with the impression that there were excess cancers among the shipyard workers, although the final conclusion of the story was that there were not. Is it any wonder that the public is misinformed?

Genetic Diseases

We have referred to "genetic defects" and "genetic disease" several times in this chapter. For those interested we here present a deeper discussion of these.[13,29]

The male sperm and the female egg which unite in the reproductive process each carry 23 threadlike chromosomes, each of which in turn is composed of many thousands of genes, and it is these genes that determine the traits of the newborn individual. At conception, the corresponding chromosomes from the two parents find one another and join, bringing the corresponding genes together. In the process, a fateful lottery takes place in which the traits of the new individual are determined by chance from a variety of possibilities passed down from previous generations. While the selections in each lottery are strictly a matter of chance, the over-all process of genetics is not, because an individual with traits less favorable for survival is less likely to reproduce. This is Darwin's celebrated law of natural selection: favorable traits are bred into the race, while unfavorable traits are bred out.

But not quite all of the entries in the lottery are inherited from previous generations, because there can be changes in them, called mutations. The information contained in genes is encoded in the structure of the complex molecules of which they are composed, so if some change occurs in this structure, the information can be changed, leading to an alteration of the traits they determine. The great majority of these altered traits are harmful, and are therefore referred to as genetic defects. Thousands of different diseases are medically recognized as arising from genetic defects. In fact an appreciable fraction of ill health, with the notable exception of infectious diseases, is due to them. These genetically related diseases range from problems that are so mild as to be hardly noticeable to some so severe as to make life impossible. In the latter category, they are responsible for something like 20% of all spontaneous abortions.

It should not be inferred that every mutation causes a genetic defect. When the chromosomes from the sperm and egg join up at conception, there is a competition between each pair of matching genes to determine which determines the trait. Some genes are dominant and others are recessive, and in the competition, the former win out over the latter.

In a small fraction of mutations, the changed gene is dominant, causing its effects to be expressed in the child with high probability. For example, achondroplasia (short-limbed dwarfism), congenital cataracts and other eye diseases, and some types of muscular dystrophy and anemia are due to this type of mutation, and it sometimes causes children to be born with an extra finger or toe (which are easily removed by cosmetic surgery shortly after birth). About a thousand different medically recognized conditions are due to dominant mutations, and roughly one percent of the population suffers from them.

But in the great majority of cases, mutations are recessive, which means that they cause characteristic diseases only in the highly unusual situation where the same mutation occurs in the corresponding genes from both parents. Sickle cell anemia, cystic fibrosis and Tay–Sachs disease are relatively well-known recessive diseases, but most of the 500–1000 known recessive diseases are extremely rare, and only about one person in 1000 suffers from them.

Another type of genetic damage involves the chromosomes rather than the genes of which they are composed. Chromosomes can be broken—recall that they are like threads—often followed by rejoining in other than their original configuration; for example, a broken-off piece of one chromosome can join to another chromosome. In some situations, there can be an entire extra chromosome or missing chromosome. About one person in 160 suffers from a disease caused by some type of chromosome aberration. Down's syndrome (mongolism) is perhaps the best-known example.

There is one other type of genetic defect that is rather different from the ones we have been discussing. About 9% of all live-born children are seriously handicapped at some time during their lives by one of a variety of diseases that tend to "run in families," like diabetes, various forms of mental retardation, and epilepsy, in which genetic factors play a role but other factors are also important. It is estimated[13] that the genetic component of these irregularly inherited diseases is something like 16%, with the other 84% due to environmental factors, food, smoking habits, etc. Effectively, then, 16% of 9%, or 1.4% of the population suffers from these diseases due to genetic factors. Adding them to the victims of dominant, recessive, and chromosomal disorders, about 3% of the population is afflicted with some type of genetic disease.

Genetic Effects of Radiation

New mutations are constantly occurring in the sex cells of people destined to bear children, but this does not cause the 3% to increase, because mutations from earlier generations are being bred out by natural selection at an equal rate. That is, there is an equilibrium between introduction of new mutations and breeding out the old mutations. The great majority of new mutations occur spontaneously as a random process, due to the random motions that characterize all matter on an atomic scale. Occasionally these motions cause a complex molecule to break apart or change its structure, and that can result in a mutation. Similar damage can be done by foreign chemicals, or viruses, or radiation which happen to penetrate into the cell nucleus where the chromosomes are housed, but here we are concerned with the last of these, the genetic effects of radiation.

It should not be inferred that a particle of radiation striking the sex cells always leads to tragedy. On an average, the nucleus of every cell in our bodies is penetrated by a particle of radiation once every three years due just to the natural radiation to which we are all exposed, but this is responsible for only one-thirtieth as much genetic disease as is caused by spontaneous mutations (that is, 1/30 of 3%, or 0.1% of the population suffers from genetic disease caused by natural radiation).

Some Calculations

In the introduction to this chapter, we discussed the number of gamma rays per second that strike an average person. To understand how these numbers are derived requires use of some definitions. One millirem (1 mrem) is defined for gamma rays as absorption by the body of 10^{-5} joules of energy per kilogram of weight. One MeV of energy is defined as 1.6×10^{-13} joules (J), and an average gamma ray from natural radiation has an energy of about 0.6 MeV, which is therefore $(0.6 \times 1.6 \times 10^{-13} =) 1 \times 10^{-13}$ J; 10^{-5} J then corresponds to 10^8 gamma rays. An average person weighs 70 kg (154 lb) whence 1 mrem of radiation exposure corresponds to being struck by $(70 \times 10^8 =) 7 \times 10^9$ gamma rays. Natural radiation exposes us to about 80 mrem/year,[3] which corresponds to $(80 \times 7 \times 10^9 \cong) 5 \times 10^{11}$ gamma rays. There are 3.2×10^7 seconds in a year, so we are struck by $(5 \times 10^{11}/ 3.2 \times 10^7 \approx) 15{,}000$ gamma rays every second. In a lifetime we are struck by about $(72 \times 5 \times 10^{11} \cong) 4 \times 10^{13} = 40$ trillion gamma rays.

A medical X-ray may expose us to 50 mrem, which would be $(50 \times 7 \times 10^9 =) 3.5 \times 10^{11}$ gamma rays. But the energy of X-rays is typically ten times lower than the 0.6 MeV we have assumed for gamma rays, so ten times as many particles would be required to deposit the same energy; this is $(3.5 \times 10^{11} \times 10 =) 3.5$ trillion X-ray particles striking us when we get an X-ray.

Since 1 mrem of gamma ray exposure corresponds to being struck by 7×10^9 particles, and the cancer risk from 1 mrem is one in 8×10^6, the risk from being struck by one gamma ray is one chance in $(7 \times 10^9 \times 8 \times 10^6 \cong) 5 \times 10^{16}$, or one chance in 50 quadrillion.

Since natural radiation exposes us to 4×10^{13} gamma rays over a lifetime, the probability that it will cause a fatal cancer is $(4 \times 10^{13}/5 \times 10^{16} \cong) 1 \times 10^{-3}$, one chance in a thousand. Our total probability of dying from cancer is now one chance in five, whence $(1/1000 \div 1/5 =) 1/200$ of all cancers are presumably caused by radiation.

REFERENCE NOTES

1. S. Novick, *The Careless Atom* (Dell Publ. Co., New York, 1969), p. 105.
2. See Appendix for this chapter.
3. National Council on Radiation Protection and Measurements (NCRP), "Natural Background Radiation in the United States," NCRP Report No. 45 (1975).
4. Calculations by the author, unpublished.
5. "Report of the President's Commission on The Accident at Three Mile Island," Washington, D.C. (1979); "Three Mile Island, A Report to the Commissioners and to the Public," Nuclear Regulatory Commission Special Inquiry Group; Ad Hoc Interagency Dose Assessment Group, "Population Dose and Health Impact of the Accident at the Three Mile Island Nuclear Station," Nuclear Regulatory Commission Document NUREG-0558 (1979). Early assessment gave an average dose of 1.7 mrem, but later revisions reduced this to 1.2 mrem.
6. James Hardin (Kentucky Department of Human Resources) private communication. He was in charge of environmental monitoring in the area.
7. *Philadelphia Evening Bulletin,* May 6, 7, 8 (1979).

8. Private communication with Health Physicists from the Ginna plant.
9. L. Garfinkel, C. E. Poindexter, and E. Silverberg, "Cancer Statistics—1980," American Cancer Society (1981).
10. United Nations Scientific Committee on Effects of Atomic Radiation (UNSCEAR), "Sources and Effects of Ionizing Radiation," United Nations, New York (1977).
11. Interagency Task Force on Ionizing Radiation, Summary of Work Group Reports, Department of HEW (1979).
12. National Council on Radiation Protection and Measurements, Radiation Exposure from Consumer Products and Miscellaneous Sources, NCRP Report No. 56, Washington, D.C. (1977).
13. National Academy of Sciences Committee on Biological Effects of Ionizing Radiation (BEIR), "The Effects on Populations of Exposure to Low Levels of Ionizing Radiation," Washington, D.C. (1980).
14. International Commission on Radiological Protection (ICRP), *Recommendations of the International Commission on Radiological Protection*, ICRP Publication No. 26 (Pergamon Press, Oxford, 1977).
15. B. L. Cohen and I. S. Lee, "A Catalog of Risks," *Health Physics*, **36**, 707 (1979).
16. R. Garrison, U.S. Department of Energy, private communication on transport accidents. Estimates for others from various sources of information.
17. C. C. Lushbaugh, S. A. Fry, C. F. Hubner, and R. C. Ricks, "Total-Body Irradiation: A Historical Review and Follow-up," in C. F. Hubner and S. A. Fry (eds.), *The Medical Basis for Radiation Accident Preparedness* (Elsevier–North Holland, Amsterdam, 1980).
18. T. F. Mancuso, A. Stewart, and G. Kneale, *Health Physics, 33*, 369 (1977).
19. E. S. Gilbert, Battelle Pacific Northwest Laboratory Document PNL-SA-6341 (1978).
20. G. B. Hutchinson, B. MacMahon, S. Jablon, C. E. Land, *Health Physics, 37*, 207 (1979).
21. U.S. General Accounting Office, "Problems in Assessing the Cancer Risks of Low-level Ionizing Radiation Exposure," Report EMD-81, Washington, D.C. (1981).
22. J. A. Reissland, "An assessment of the Mancuso study," Publication NRPB-79, U.K. National Radiological Protection Board, Didcot, Berk. (1978); T. W. Anderson, *Health Physics, 35*, 743 (1978); A. Brodsky, testimony before the Subcommittee on Health and the Environment, U.S. House of Representatives, Washington, D.C., 8 February 1978; B. L. Cohen, *Health Physics, 35*, 582 (1978); S. M. Gertz, *ibid., 35*, 723 (1978); E. S. Gilbert, "Methods of Analyzing Mortality of Workers Exposed to Low Levels of Ionizing Radiation," Report BNWL-SA-634, Battelle Pacific Northwest Laboratory, Richland, Washington, May (1977); E. Gilbert and S. Marks, *Health Physics, 37*, 791 (1979); *ibid., 40*, 125 (1981); J. W. Gofman, *ibid., 37*, 617 (1979); D. J. Kleitman, "Critique of Mancuso–Stewart–Kneale report" (prepared for the U.S. Nuclear Regulatory Commission, Washington, D.C., 1978); S. Marks, E. S. Gilbert, and B. D. Breiten-

stein, "Cancer mortality in Hanford workers," Document IAEA-SM-224, International Atomic Energy Agency, Vienna (1978); R. Mole, *Lancet*, **i**, 582 (1978); "Staff Committee Report of November 1976," Nuclear Regulatory Commission, Washington, D.C. (1976); "Staff Committee Report of May 1978," Nuclear Regulatory Commission, Washington, D.C. (1978); "The Windscale Inquiry," Her Majesty's Stationery Office, London (1978); D. Rubenstein, "Report to the U.S. Nuclear Regulatory Commission," Nuclear Regulatory Commission, Washington, D.C. (1978); L. A. Sagan, "Low-Level Radiation Effects: The Mancuso Study," Electric Power Research Institute, Palo Alto, California (1978); B. S. Sanders, *Health Physics*, **34**, 521 (1978); F. W. Spiers, *ibid.*, **37**, 784 (1979); G. W. C. Tait, *ibid.*, **37**, 251 (1979).

23. B. L. Cohen, "The Low-Level Radiation Link to Cancer of the Pancreas," *Health Physics*, **38**, 712 (1980).

24. I. D. J. Bross, M. Ball, and S. Falen, *American Journal of Public Health*, **69**, 130 (1979).

25. J. D. Boice and C. E. Land, *American Journal of Public Health*, **69**, 137 (1979).

26. T. D. Luckey, *Hormesis with Ionizing Radiation* (CRC Press, Boca Raton, Fla. 1980).

27. A. F. Cohen and B. L. Cohen, "Tests of the Linearity Assumption in the Dose–Effect Relationship for Radiation Induced Cancer," *Health Physics*, **38**, 53 (1980).

28. National Academy of Sciences Committee on Biological Effects of Ionizing Radiation (BEIR), "The Effects on Populations of Exposure to Low Levels of Ionizing Radiation," Washington, D.C. (1972).

29. B. L. Cohen, "Perspective on Genetic Effects of Radiation," *Health Physics* (in press).

30. E. B. Hook, "Rates of Chromosome Abnormalities at Different Maternal Ages," *Obstetrics and Gynecology*, **58**, 282 (1981).

31. J. M. Friedman, "Genetic Disease in the Offspring of Older Fathers," *Obstetrics and Gynecology*, **57**, 745 (1981).

32. W. J. Schull, M. Otako, and J. V. Neel, "Genetic Effects of Atomic Bombs: A Reappraisal," *Science*, **213**, 1220 (1981).

33. R. J. Lewis (ed.), "Registry of Toxic Effects of Chemical Substances," U.S. Public Health Service, November (1981) (available by computer access).

34. L. Fishbein, in *Chemical Mutagens*, Vol. 4, A. Hollaender (ed.) (Plenum Publ. Co., New York, 1976) pp. 219ff.

35. K. Sax and H. J. Sax, "Radiomimetric Beverages, Drugs, and Mutagens," *Proceedings of the National Academy of Sciences*, **55**, 1431 (1966).

36. L. Ehrenberg, G. von Ehrenstein, and A. Hedgran, "Gonad Temperature and Spontaneous Mutation Rate in Man," *Nature*, December 2, 1433 (1957).

37. U.S. Department of HEW, "Antenatal Diagnosis," National Institutes of Health Publication No. 79-1973 (1979).

38. G. W. Beebe, H. Kato, and C. E. Land, "Mortality Experience of Atomic Bomb Survivors 1950–1974," Radiation Effects Research Foundation Technical Report RERF TR 1-77 (1977). The data for Hiroshima and Nagasaki were added here.

39. National Council on Radiation Protection and Measurements (NCRP), "Review of NCRP Radiation Dose Limit for Embryo and Fetus in Occupationally Exposed Women," NCRP Report No. 53 (1977).
40. W. J. Blot and R. W. Miller, "Mental Retardation Following In Utero Exposure to the Atomic Bombs," *Radiology*, **106**, 617 (1973).
41. G. B. Avery, L. Menesis, and A. Lodge, "The Clinical Significance of Measurement of Microcephaly," *American Journal of Diseases of Children*, **123**, 214 (1972).
42. A. Stewart *et al.*, "Malignant Disease in Childhood and Diagnostic Irradiation in Utero," *Lancet*, **ii**, 447 (1956); A. Stewart and G. W. Kneale, "Radiation Dose Effects in Relation to Obstetric X-Rays and Childhood Cancer," *Lancet*, **i**, 1185 (1970).
43. B. MacMahon, "Prenatal X-Ray Exposure and Childhood Cancer," *Journal of the National Cancer Institute*, **28**, 1173 (1962).
44. S. Jablon and H. Kato, "Childhood Cancer in Relation to Prenatal Exposure to Atomic Bomb Radiation," *Lancet*, **ii**, 1000 (1970).
45. B. E. Oppenheim, M. L. Griem, and P. Meier, "Effects of Low Dose Prenatal Irradiation in Humans," *Radiation Research*, **57**, 508 (1974).
46. Rashed Fawwaz, Columbia-Presbyterian Hospital, private communication (1982).
47. A. K. Poznanski, C. Kanellisas, D. W. Roloff, and R. C. Borer, "Radiation Exposure to Personnel in a Neo-natal Facility," *Pediatrics*, **54**, 139 (1974).
48. David Milstein, Albert Einstein Medical Center, private communication (1982).
49. B. L. Cohen, "The Cancer Risk from Low Level Radiation," *Health Physics*, **39**, 659 (1980). Many references are given therein.
50. E. W. Webster, "Cancer Risks from Low Level Radiation—A Commentary on the BEIR Report 1980," in Critical Issues in Setting Radiation Dose Limits, Proceedings No. 3, National Council on Radiation Protection and Measurements (NCRP) (1982).
51. A. M. Kellerer and H. H. Rossi, *Current Topics in Radiation Research*, **8**, 85 (1972).
52. D. C. Lloyd *et al.*, "The Relationship Between Chromosome Aberrations and Low LET Radiation Dose to Human Lymphocytes," *International Journal of Radiation Biology*, **28**, 75 (1975).
53. E. J. Purrott and E. Reeder, "The Effects of Changes in Dose Rate on the Yield of Chromosome Aberrations in Human Lymphocytes Exposed to Gamma Radiation," *Mutation Research*, **35**, 437 (1976).
54. M. Terzaghi and J. B. Little, "Radiation Induced Transformation in a C3H Mouse Embryo-Derived Cell Line," *Cancer Research*, **36**, 1367 (1976).
55. A. Han and M. M. Elkind, "Transformation of Mouse C3H Cells by Single and Fractionated Doses of X-Rays and Fission Spectrum Neutrons," *Cancer Research*, **39**, 123 (1979).
56. R. L. Ullrich *et al.*, "The Influence of Dose and Dose Rate on the Incidence of Neoplastic Disease in RFM Mice after Neutron Irradiation," *Radiation Research*, **68**, 115 (1976).

57. M. P. Finkel and B. O. Biskis, *Progress in Experimental Tumor Research*, **10**, 72 (1968).
58. R. E. Rowland, A. F. Stehney, and H. F. Lucas, *Radiation Research*, **76**, 368 (1978).
59. B. L. Cohen, "Failures and Critique of the BEIR-III Lung Cancer Risk Estimates," *Health Physics*, **42**, 267 (1982).
60. T. F. Dougherty and C. W. Mays, "Bone Cancer Induced by Internally Deposited Emitters in Beagles," in *Radiation Induced Cancer*, p. 361, International Atomic Energy Agency, Vienna (1969).
61. C. W. Mays, H. Spiess, and A. Gerspach, "Skeletal Effects Following ^{224}Ra Injections into Humans," *Health Physics*, **35**, 83 (1978).
62. T. Kitabatake, T. Watanabe, and S. Koga, *Strahlentherapie*, **146**, 599 (1973).
63. R. D. Evans, "Radium in Man," *Health Physics*, **27**, 497 (1974).
64. J. B. Storer, "Radiation Carcinogenesis," in *Cancer*, F. F. Becher (ed.), Plenum Press, New York (1975). This is a review that lists references to the various animal studies.
65. H. Nishiyama *et al.*, *Cancer*, **32**, 1301 (1973).
66. T. Najarian, *Boston Globe*, Feb. 19, 1978.
67. T. Najarian and T. Colton, *Lancet*, **i,** 1018 (1978).
68. R. A. Rensky *et al.*, "Cancer Mortality at a Naval Nuclear Shipyard," *The Lancet*, 31 Jan. 1981, p. 231.
69. S. Wermiehl, "Doctors Shift on Shipyard: Kennedy Chides Portsmouth Researcher," *Boston Globe*, June 20, 1979.

Chapter 3 / THE FEARSOME REACTOR MELTDOWN ACCIDENT*

"Meltdown" has become a household word, referring to an accident that has never occurred, a largescale melting of the fuel in a nuclear power reactor. The media frequently refer to it as "the ultimate disaster," and many people envision it to be accompanied by stacks of dead bodies and a devastated landscape, much like the aftermath of a nuclear bomb attack.

---

* Nearly all of my knowledge and experience has been confined to the "pressurized water reactor" (PWR), manufactured in the United States by Westinghouse, Babcock and Wilcox, and Combustion Engineering. About 70% of the reactors in this country are of that type. The discussion here generally refers to them.

All but one of the other 30% are "boiling water reactors" (BWR) manufactured by General Electric. Their safety problems and defenses are similar in general but rather different in detail, and I am not sufficiently familiar with them to discuss them here. They will be mentioned only in a few situations where they have attracted public attention.

Most of the field of reactor accidents and safety is somewhat removed from my principal areas of expertise. The latter include what happens after radioactivity is released to the environment in accidents, but on causes of accidents and defenses against them, which is the heart of the subject, I have produced only one research paper. This chapter should therefore be considered less authoritative than Chapters 2, 4, 5, 6, and 7. However, I have done a considerable amount of reading and discussing with experts in preparing it.

On the other hand, the two principal reports on the Three Mile Island accident[1,2] agree that even if there had been a meltdown in that reactor, there very probably would have been essentially *no* harm to human health and no environmental damage beyond the plant boundary. I know of no technical reports that have claimed otherwise. Moreover, all scientific studies agree that in the great majority of meltdown accidents there would be no detectable effects on human health, immediately or eventually. According to the official government estimate, there would have to be a meltdown every two weeks somewhere in the United States before nuclear power would be as dangerous as coal burning. Once again we have here a vast gulf of public misunderstanding and a grossly distorted perspective.

WAS THREE MILE ISLAND A NEAR MISS TO DISASTER?

The principal reason for the discrepancy between the public's impressions and the technical analyses is that nuclear reactors are sealed inside a very powerfully built structure called the "containment," which under ordinary circumstances would prevent the escape of radioactivity even if the reactor fuel were to melt. A typical containment[3-5] is constructed of three-foot-thick concrete walls heavily reinforced by thick steel rods (cf. Fig. 10) welded into a tight net around which the concrete is poured—in fact there is so much steel reinforcing that special techniques had to be developed to get the concrete to become distributed around it as it is poured. In addition, the inside of the containment is lined with thick steel plate welded to form a tight chamber which can withstand very high internal pressure.

The containment provides a broad range of protection for the reactor against external forces, such as a tornado hurling an automobile, a tree, or a house against it, an airplane flying into it, or a large charge of chemical explosive detonated against it. In a meltdown accident, however, the function of the containment is to hold the radioactive material inside. Actually it need only do this for several hours, because there are systems inside the containment for removing the radioactivity from the atmosphere. One type blows the air through filters in an operation similar in principle to that of household vacuum cleaners.* In another, water sprinklers remove the dust from the air. There are charcoal filter beds or chemical sprays for removing certain types of airborne radioactivity. Most radioactive materials, however, would simply

---

* The radioactive dust is trapped in filters which are then disposed of as described in Chapter 6.

Fig. 10. A construction worker on the steel-rod-reinforced containment structure of a Westinghouse reactor.

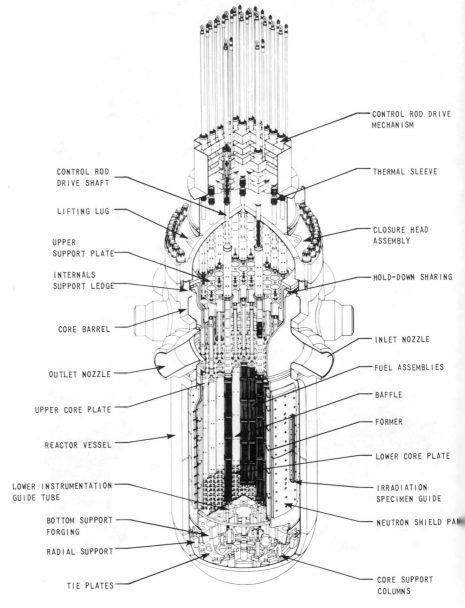

Fig. 11. A Westinghouse nuclear reactor.

get stuck to the walls of the building and the equipment inside, and thereby be removed from the air. Thus, if the containment holds even for several hours, the health consequences of a meltdown would be greatly mitigated. In the Three Mile Island accident, there was no threat to the containment. The investigations have therefore concluded that even if there had been a meltdown, the containment would very probably have prevented the escape of any large amount of radioactivity.[1,2] In short, even if the Three Mile Island accident was a "near miss" to a meltdown (a highly debatable point we will discuss later), it was definitely *not* a near miss to a health disaster. Although that point was made in reports[1,2] that have received very wide media coverage, this coverage has rarely included that particular point.

Much of the public consequently continues to believe wrongly that Three Mile Island was a near miss to a disaster. With all the continued coverage that accident continues to receive, including TV specials on each anniversary, there have been numerous opportunities to straighten out that matter, but somehow the media people don't seem to think that that is important. I have challenged them on this point face to face, and all I get is a shoulder shrug with a comment like "maybe we should." I am personally convinced that they don't tell the public it was not a near miss because that would tend to ruin their big story.

ROADS TO MELTDOWN

In order to understand the meltdown accident, we must go back to its origins. A nuclear power reactor is basically just a water heater, evolving heat from fission processes in the fuel. This heats the water surrounding the fuel (cf. Fig. 11) and the hot water is used to produce steam. The steam is then employed as in coal- or oil-fired power plants to drive a turbine which turns a generator (sometimes called a "dynamo") which produces electric power (cf. Fig. 12). There are features of the nuclear water heater that differentiate it from water heaters in our basements or the coal- or oil-fired boilers that produce steam for various purposes in industrial plants. First, the waste products from the burning do not go up a chimney or settle to the bottom as an ash, but rather are retained inside the fuel, which normally maintains its original size and shape—nuclear fuel does not crumble into ashes or get converted into a gas as do coal and oil fuels. Second, these waste products are radioactive, which means that they emit radiation. Third, because of their radioactivity, these wastes continue to heat the fuel even after the

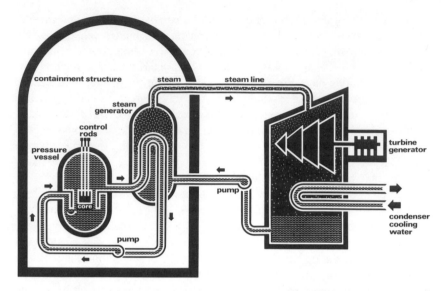

Fig. 12. Diagram (highly simplified) of a pressurized water reactor power plant. Water is heated to 600°F by energy released in fission reactions in the reactor (it is prevented from boiling by maintaining high pressure), and pumped to the steam generator, where its heat is transferred to a secondary water system. The water in the latter is thereby boiled to become steam, which drives the turbines. The turbine drives the generator, which produces electricity. It is necessary to condense the steam into water, greatly reducing its pressure, after it exits the turbine—otherwise there would be no tendency for the steam to rush through the turbine and thereby cause it to turn. The steam is condensed in the condenser by cooling it with water brought in from some outside source. The water formed by condensation of steam is pumped back to the steam generator to be reused.

reactor is shut down[6]; it is therefore necessary to continue to provide some water to carry this heat away.

If, for some reason, there is no water available to remove this heat—this is called a loss of coolant accident (LOCA)—the fuel will heat up and will eventually melt—this is called a "meltdown." A meltdown releases the radioactivity sealed inside. Some of this radioactivity would come off as airborne dust which has a potential for damaging public health if it is released into the environment. If there is some water in the reactor but not enough, the situation may be even worse because steam reacts chemically with the fuel casing material (an alloy of zirconium) at high temperature (2700°F) releasing hydrogen, an inflammable and potentially explosive gas, and providing additional heat, thereby accelerating the fuel melting process.

In most scenarios that have been concocted by reactor safety analysts,[7] a LOCA would be started by a large pipe suddenly breaking open, allowing the reactor cooling water to escape. This has never happened, but we will consider it in some detail later in this section. In the Three Mile Island accident,[8] the LOCA occurred as a result of a valve failing to close while the operators were led to believe that it was closed because they misinterpreted their instrument readings. According to one estimate,[2] a meltdown would have occurred if the water had continued to escape through the open valve for another 30 to 60 minutes.

How close was Three Mile Island to a meltdown? There were many unusual aspects to the instrument readings at the time. Clearly something very strange was going on. A number of knowledgeable people were therefore trying to figure out what to do when one suggested closing an auxiliary valve in the pipe through which water was escaping. Within less than a minute after it was closed, a telephone call came in from another expert working at home asking whether this auxiliary valve was closed,[2] so it cannot be claimed that a meltdown was prevented by the luck of one man's recognizing the right thing to do. It is difficult to prove that, if neither of the two had thought of closing the valve, someone else would have, but there were a lot of people involved in analyzing the information, and there would have been further clues developing before a meltdown would have occurred. Some analyses indicate that there would not have been a meltdown even if the valve had not been closed, as there was a small amount of water still being pumped in. Moreover, there were other actions that would have prevented a meltdown, like allowing emergency cooling water to enter the reactor—it was purposely shut off* because the operators' false interpretation of their instrument readings led them to believe that there was *too much* water in the reactor.

In any case, the widely publicized statement that the Three Mile Island accident came within thirty to sixty minutes of a meltdown seems to be sufficient to achieve the effect desired by those who continually use it to scare the public. I often wonder why this works—when we drive on a high speed highway, on every curve we are within a *few seconds* of being killed if nothing is done—that is, if the steering wheel is not turned at the proper time. And don't forget that even if a meltdown had occurred, there very probably would have been no health consequences, since the radioactivity would have been contained.

---

* More accurately, it was throttled down to allow only about 10% of the normal water flow.

As a result of the Three Mile Island accident[9] there have been great improvements in instrumentation, information availability to the operators, and operator training, and there is now a requirement that a graduate engineer be on hand at all times, so there will probably never again be a LOCA arising from faulty interpretation of instrument readings. With that road to a meltdown now blocked, we can go back to what was always believed to be the most probable road, a break in the system allowing the water to escape. Since the temperature and pressure are normally very high (typically 600°F and 2200 lb/in.^{2}), the water would come shooting out of the break as steam in a process picturesquely called "blowdown." The first line of defense against a melt-down—or a blowdown—is a series of measures to prevent the system from breaking open in this way.[10,3]

One of these measures is very elaborate quality control on materials and workmanship, far superior to that in any other industry. No effort or expense is spared in choosing the highest quality materials and equipment, nor in requiring the most demanding specifications for safety-related parts of the system. The second measure is a very elaborate inspection program, including X-ray inspection of every weld, and other inspections with magnetic particle and ultrasonic techniques during construction, followed by periodic ultrasonic and visual inspections after the reactor has gone into operation. The visual inspection program, for example, includes removal of insulation from pipes to search for imperfections or signs of cracking. One problem originally discovered by these inspections, "corrosion cracking," is discussed in the last section of this chapter. A third measure is a variety of leak detection systems: ordinarily a large break starts out as a small crack which allows some of the water and the radioactivity it contains to leak out. Leaking water becomes steam as it emerges (its temperature is close to 600°F), increasing the humidity; there are instruments installed to detect this increased humidity. Much of the radioactive material emerging with the leaking water attaches to airborne dust, and there are instruments in place for detecting increased radioactivity in this dust. These systems for detecting increased humidity and increased radio-activity in dust act as sensitive indicators for leaks, therefore serving as early warnings of possible cracks in the system.

Thanks to these precautions there has never been a cracking open of a reactor system leading to a loss of coolant accident (LOCA). But if there should be, a second line of defense is ready and waiting. It is the emergency core cooling system (ECCS), an elaborate complex of systems to pour water back into the reactor if the regular cooling water were lost in a blowdown. There are several separate pumping systems for doing this, one of which would provide sufficient water to save the reactor in most cases—in all cases, two would do the job.[11] More details on the ECCS are given in the Appendix. To guard against failure of electric power to drive these pumps, two separate power lines are brought into the plant from off-site sources, and there are several diesel generators in the plant, any one of which could provide the needed electricity. There is thus only a very small probability that these systems would fail to deliver the needed water following a LOCA.

When the first of this water from the ECCS would reach the hot fuel, it would flash into steam, and at one time there was some concern as to whether this might prevent further water from reaching and cooling the fuel. Some of the first tests of small mock-ups, performed in 1970–1971, indicated that this might be the case. The problem thus received very wide publicity.[12] This was the situation that brought the antinuclear Union of Concerned Scientists (UCS) into prominence, as they asked for a halt to reactor licensing until the problem was resolved.[13] As a result, a series of hearings extending over a year was held in Washington in 1972–1973. As might be expected, the TV cameras ground away when UCS witnesses were testifying, but were nowhere in evidence when the other viewpoint was presented. At least some segments of the public became engrossed in the controversy. I have encountered people ranging from housewives to engineers who were eager to tell me in some technical detail why the ECCS would never work.

As a result of the hearings, changes were introduced in reactor operation as a temporary measure to reduce the performance required of the ECCS if a LOCA should occur, and a crash research program costing hundreds of millions of dollars was instigated to settle the unresolved questions. As more sophisticated experimental tests and computer analyses were developed, it became increasingly clear in the 1975–1978 time period that the ECCS would work. There were over 50 tests, far more realistic and sophisticated than the 1971 tests, and all came out favorably. The question was finally resolved in 1978 when a test reactor specifically designed to test the ECCS (called LOFT, for loss of fluid test) came into operation at the Idaho Nuclear Engineering laboratory and was put through various types of LOCAs. In all cases, the ECCS performed better than had been estimated.[14] For example, in the first LOFT test, the best estimate from the computer analysis was a maximum temperature of 1376°F, the conservative calculation used for the safety analysis gave 2018°F, but the highest measured temperature was only 960°F. In the second LOFT test, carried out under rather different conditions, these temperatures were 1360, 2205, and 1185°F, respectively. These examples also demonstrate how conservative estimates rather than "best estimates" are generally used in safety analyses. This is good engineering practice, but it is not usually recognized by those who use such estimates to frighten the public.

In view of all the media publicity at the time the question was raised, one might have expected at least some media coverage for its final resolution, but if there was any, I missed it. I am still occasionally told that the ECCS won't work. When I point out the results of the LOFT tests, the inevitable reply is that the questioner never heard about them. At least as recently as 1979, the Union of Concerned Scientists, which originally came to prominence with claims that the ECCS would not work, continued to send out brochures[15] stating that tests of the ECCS had shown it to be an unmitigated failure. The last test that could possibly be interpreted that way was in 1971; they chose to ignore the dozens of much better tests from 1973 to 1978, and even the first few definitive LOFT tests which had been completed by that time.

One type of LOCA in which the ECCS would not prevent a meltdown is a large crack in the *bottom* of the reactor vessel, since water injected by the ECCS would simply pour out through that crack—this is not a problem with pipe breaks since all significant pipes enter the vessel near its top. This problem was intensively investigated by the British as part of their decision to convert from their own type to American-type reactors and they concluded[16] that, in view of the large thicknesses (see Fig. 11) and high quality of the materials used, the probability of a large crack in the reactor vessel is so small as to be negligible. There is also an elaborate inspection program[4] to ensure that the high quality of the reactor vessel material is maintained. One potential problem in this regard, "pressurized thermal shock," has received widespread publicity. It is discussed later in this chapter.

While every effort is being made to block the roads to meltdown, there is always a possibility of a road being opened by successive failures in the various lines of defense we have described. Or perhaps there is some obscure road to meltdown that no one has ever thought of in spite of the thousands of man-years of technical effort on this problem. If nuclear power becomes a flourishing industry, there probably will be meltdowns somewhere some day. But if and when they occur, there is still one final line of defense—the containment—which should protect the public from harm in most cases. Let's now consider the reliability of that line of defense.

How Secure Is the Containment?

If the reactor system breaks open, the water and steam inside pours out into the containment building. When water is pumped in by the emergency core cooling system, some of it overflows, and when it surrounds the fuel it boils into steam which goes out through the break into the containment. We thus expect the containment to be filled with steam, with a lot of excess water on the floor. This is true in nearly all potential loss of coolant accidents, even if the system does not break open, as was the case in the Three Mile Island accident. In addition, heat is being fed into this water and steam by the radioactivity in the fuel, by chemical reactions of steam with the fuel casing, and by burning of the hydrogen generated in those reactions. The most important threat to the security of the containment is that this heat will raise the pressure of the steam to the point where it will exceed the holding power of the containment walls, about ten times normal atmospheric pressure.

In order to counteract this threat, there are systems for cooling the containment atmosphere.[4,17] One such system sprays cool water into the air, a very efficient way of condensing steam; when it exhausts its stored water supply, it picks up water from the containment floor, cools it, and then sprays it into the air inside the containment. Another type of system consists of fans blowing containment air over tubes through which cool water is circulating. There are typically five of these systems but only one (or in rare cases, two) need be operable in order to assure that the containment is adequately cooled. In most cases, one of the systems is driven by a diesel engine so as to be available in the event of an electric power failure. A more quantitative treatment of the containment cooling problem is given in the Appendix.

Since they are safety related, these systems are subject to elaborate quality control in fabrication and are frequently inspected and tested, so it seems reasonable to expect at least most of these systems to function properly if an accident should occur. All of them were functional during the Three Mile Island accident, and that is why it is concluded[1,2] that the containment would have prevented the escape of radioactivity even if there had been a meltdown there.

Two other possible mechanisms for breaking open the containment have been discussed. One of these is a steam explosion, which has received considerable research attention[17] and was given wide publicity in the fictional movie "The China Syndrome." The worst situation is in a meltdown where the molten fuel falls into a pool of water at the bottom of the reactor vessel producing so much steam so suddenly that the top of the reactor vessel would be blown off and be hurled upward with so much force that it would break open the top of the containment building. To obtain a powerful enough explosion to do this, the hundred tons of molten fuel would have to be divided into drops the size of necklace beads, with half of them contacting the water simultaneously within a half second! This would seem to be an obscure possibility, as it would be much more likely for the fuel to drop into the water in rather large chunks over a period of many seconds or minutes. As a result of recent more elaborate studies, the steam explosion has now been essentially discounted by most analysts as a possible way of breaking the containment.

In "The China Syndrome" it is implied that a sufficiently powerful explosion can occur when the molten fuel melts its way into the ground and comes into contact with ground water, but this is obviously unrealistic since this would be an infinitely more gradual contact than the one described above. A fictional movie need not be realistic, of course, but it is important for the audience to recognize that point.

The movie also makes an issue of groundwater contamination following a meltdown accident. However, when molten fuel would first come into contact with groundwater, the latter would flash into steam which would build up a pressure to keep the rest of the groundwater away. There would thus be little contact until the molten fuel cooled and solidified many days later. It would then be in the form of a glassy mass that would be highly insoluble in water, so there would be relatively little groundwater contamination. If that were judged to be a problem, there would be plenty of time to construct barriers to permanently isolate the radioactivity from groundwater. It is difficult to imagine a situation in which there would be any adverse health effects from groundwater contamination.

The other possible mechanism for breaking the containment is a hydrogen explosion which has received substantial research attention[17] and achieved notoriety in the Three Mile Island accident. The consensus of the research seems to be that even if all the hydrogen that could be generated in an accident were to explode at once, the forces would not be powerful enough to break most containments, including the one at Three Mile Island.[18] Moreover, the hydrogen would be produced gradually and there are many sources of sparks (e.g., electric motors) which would cause it to burn in a series of fires and/or small explosions not nearly large enough to threaten the containment. There was such a small explosion in the Three Mile Island accident.

Three of the 48 U.S. pressurized water reactor (PWR) containments store large volumes of ice inside to reduce steam pressure in an accident. Since the presence of ice is a failproof method for cooling the surroundings and thereby avoiding high steam pressure, it was not considered necessary to build the containment walls so powerfully or to make the containment volumes so large. These containments are more vulnerable to a hydrogen explosion,[18] and are therefore fitted with numerous gadgets for generating sparks to be extra certain that hydrogen ignites before large quantities can accumulate.[19]

The boiling water reactor (BWR) containments are very much smaller in volume than those of pressurized water reactors (PWR); hence, they are much more vulnerable to pressures generated by a hydrogen explosion. It is therefore required that they be operated with an inert gas, rather than air, in the containment.[18,20] There is thus no oxygen with which the hydrogen can combine, so there can be no hydrogen explosion.

There are still things not understood about hydrogen explosions, and there is a great deal of research being devoted to that subject,[21] but all indications at present are that they are not a serious problem. If new findings should alter this conclusion, there are protective measures available to reduce the hazards.

Of course explosions inside the containment, even if they do not crack the walls, can damage equipment and this can cause problems. For example, if they disabled all of the heat removal systems, the containment might be broken by steam pressure. However, the probability for disabling many separate systems would be very small.

The systems we have described in this section and the previous one for averting a catastrophic accident constitute a "defense in depth," which is the guiding principle in designs for reactor safety. If the quality assurance fails, the inspections ordinarily provide safety. If the inspection programs fail, the leak detection saves the day. If that fails, the ECCS protects the system. And if the ECCS fails, the containment averts damage to the public. Moreover, each of these systems is itself a defense in depth; for example, if one of the ECCS water injection systems fails, another can do its job, and if both fail a third can provide sufficient water.

One sometimes hears statements to the effect that reactors are safe if everything goes right, but if any piece of equipment fails or if an operator makes a mistake, disaster will result. This statement is absolutely WRONG. In reactor design it is assumed that all sorts of things will go wrong—pipes will break, valves will stick, motors will fail, operators will push the wrong button, etc., etc., but there is "defense in depth" to cover these malfunctions or series of successive malfunctions.

Of course the depth of the defense is not infinite. If each line of defense would crumble, one after the other, there could be a disaster. But as the depth of the defense is increased, the probability for this to happen is rapidly decreased. If each line of defense has a chance of failure equal to that of drawing the ace of spades out of a deck of shuffled cards—one chance in 52, the probability for five successive lines of defense to fail is like the chance of drawing the ace of spades successively out of five decks of well-shuffled cards—one chance in $52 \times 52 \times 52 \times 52 \times 52$, or one chance in 380 million!

There have been cases where one of the lines of defense has failed in nuclear power plants. Utilities have been heavily fined by NRC for such things as leaving a valve closed and thereby compromising the effectiveness of one of the emergency systems. These incidents are often given publicity as failures that could lead to a meltdown. But the media coverage rarely bothers to point out that there are several lines of defense remaining unbreached between these events and a meltdown—not to mention that there is still a major line of defense, the containment, remaining even if a meltdown occurs. Of course bringing out those points would detract from their story—usual media style is instead to follow their description of the incident with a reminder to the audience that a meltdown is the "ultimate disaster."

THE PROBABILITIES

In considering the hazards of a reactor meltdown accident, once again we find ourselves involved in a game of chance governed by the laws of probability. By setting up additional lines of defense, or by improving the ones we now have, we can reduce the probability of a major accident, but we can never reduce it to zero. This should not necessarily be discomforting since we already are engaged in innumerable other games of chance with disastrous consequences if we lose—earthquakes, tornadoes, fires, war, famine, disease epidemics, toxic chemical releases, and dam failures, to name a few. In fact, participating in this new game of chance may save us from participating in others brought on by alternative actions, and it may therefore reduce our total risk: building a nuclear power plant may remove the need for a hydroelectric dam whose failure can cause a disaster, or for a coal-burning power plant whose air pollution might be disastrous. The important question is: what is the *probability* of a disastrous meltdown accident?

Several studies have been undertaken to answer this question. The best known of these was sponsored by the U.S. Nuclear Regulatory Commission (NRC) and directed by Dr. Norman Rasmussen, an MIT professor.[7] It extended over several years, involved many dozens of scientists and engineers, costing over $4 million before its final report was issued in 1975. The report bore the document designation "WASH-1400" and was titled "Reactor Safety Study," but it is best known as the "Rasmussen Study." It used a method known as "fault tree analysis" which is discussed in the Appendix. Its history does not stop in 1975. The Union of Concerned Scientists (UCS) published a critique of it[22] with its own probabilities in 1977 and we will quote some of its conclusions. An independent review chaired by Professor Harold Lewis of the University of California was commissioned by NRC and reported[23] in 1978. The principal finding of the Lewis panel was that the uncertainties in the probabilities given by the 1975 Rasmussen Study were larger than originally stated, but that there is no reason to believe that the probabilities were either too large or too small. The Lewis panel also took exception to the 12-page Executive Summary issued with the Rasmussen Study. The NRC accepted the Lewis panel report in 1979 so in our references to the Rasmussen Study we will not use either the Executive Summary or the uncertainty estimates. There have been other, more recent, reactor safety studies, including one by the West German government and one by the Swedish government; they use a rather similar methodology and obtain results rather similar to those of the Rasmussen Study after adjustments for differences in distribution of population.

There have been some new scientific and technical developments which substantially reduce the danger from that projected by the Rasmussen Study. Rasmussen gives a 30% probability for the containment eventually to be ruptured or bypassed following a meltdown, much larger than seems reasonable from our discussion above and in the Appendix. This is partly because Rasmussen gave a large weight to the steam explosion, which no safety analysts now accept as a credible means of breaking the containment. In addition, Rasmussen found a large probability for the containment to be bypassed by the failure of two valves, and as a result that problem has largely been remedied. A more modern probability estimate for the containment to be ruptured or bypassed is 1%–3%.[24]

On the other hand, there has been new information tending to indicate that meltdowns are more likely than estimated by the Rasmussen Study. A number of problems omitted in that study have come to light under further analysis, or as a result of operating experiences in reactors. For example, there is now more appreciation of possible consequences of flooding and earthquakes, design flaws have been found in a few plants, some of the hardware has been found to be less reliable than had been expected, and detailed studies of individual plants have shown that they have particular weaknesses not considered in the study. These things taken together would seem to indicate that the Rasmussen team underestimates the probability of a meltdown up to ten-fold. This is about the maximum underestimate possible in view of operating experience, as will be discussed below. The overall picture seems to be that meltdowns are several times more likely, but containment failures following them are several times less likely than estimated by the Rasmussen Study. These two changes roughly compensate, leaving the probability of radioactivity releases to the environment more or less unchanged.

But even more important is new information on the chemical behavior of the radioactive material released in a meltdown. Some elements, most importantly, iodine, but also cesium, tellurium, rubidium, strontium, and barium, have chemical forms that are quite volatile. At the time of the Rasmussen Study, there was little information on what chemical forms to expect in a meltdown accident, so rather than risk underestimating the danger, it was assumed that these elements would be mostly in their volatile forms. Suspicions were aroused when it was found that the amount of radioactive iodine released in the Three Mile Island accident was thousands of times less than what would be expected under that assumption. Investigations were therefore undertaken and it was found[25] that in all situations where a great deal of water is present, these potentially volatile elements take forms that dissolve in water and become nonvolatile. A close analogy is chlorine, which is very abundant in sea water. Chlorine gas is extremely volatile and poisonous—it was used as a poison gas in World War I. If the chlorine in sea water were in this form, it would be impossible to live near the seashore—Boston, New York, Miami, New Orleans, Los Angeles, San Francisco, etc. would be uninhabitable. But fortunately the chlorine in sea water is in the form of sodium chloride, the table salt we put in our food, which is not volatile. Iodine is chemically similar to chlorine. There is now a great deal of experimental evidence that, under accident conditions, it is converted into chemical forms similar to table salt. Similarly favorable results were found for the other elements that had been assumed to be volatile.

These findings are rather new and it will take some time before they will be accepted by the Nuclear Regulatory Commission for use in licensing and risk analysis. We will therefore use the results of the Rasmussen Study in our discussion, although it is likely that they will be reduced in the not-too-distant future.

The Rasmussen Study[7] estimates that a reactor meltdown may be expected about once every 20,000 years of reactor operation, while the report by the antinuclear activist UCS[21] estimates once every 2000 years. There have been well over 1000 years of applicable commercial reactor operation* throughout the World, plus about 3000 years of naval reactor operation* in which the problems are quite similar, all without a meltdown. If the UCS estimate is correct, we are very lucky not to have had a meltdown by now,

---

* For example, 100 reactors operating for one year is 100 years of reactor operation. Since accidents are largely random events, unrelated to the age of a plant (the Three Mile Island accident occurred in a plant that had only been operating for a few months), the number of accidents should be proportional to the number of reactor-years. Learning from experience should reduce the number of accidents per reactor year, but that is not taken into account here.

whereas the Rasmussen estimate is that there is about a 20% probability that there would have been one.

The Rasmussen estimate for the frequency of a loss of coolant accident (LOCA) is one every 2000 years of reactor operation. After more than 4000 years of combined civilian and military reactor operation there has been only one LOCA—Three Mile Island. The Rasmussen Study was therefore not overoptimistic on that probability.

We now turn to the consequences of a meltdown. Since it gives more detail, we will quote the results of the Rasmussen Study here; the antinuclear activist UCS viewpoint[22] can be roughly interpreted as multiplying all consequences by a factor of 10.

In most meltdowns the containment is expected to maintain its integrity for a long time, so the number of fatalities should be zero. In one out of five meltdowns there would be over 1,000 deaths,* in one out of a hundred there would be over 10,000 deaths, and in one out of 100,000 meltdowns, we would approach 50,000 deaths (the number we get each year from motor vehicle accidents). Considering all types, we expect an average of 400 fatalities per meltdown; the UCS estimate[22] is 5000. Since air pollution from coal burning is estimated to be causing 10,000 deaths each year in the United States,[26] for nuclear power to be as dangerous as coal burning there would have to be 25 meltdowns per year (10,000 ÷ 400 = 25), or one meltdown every two weeks somewhere in the United States according to the Rasmussen Study; according to the antinuclear activist UCS, there would have to be a meltdown every six months. Since there has never been a single meltdown, clearly we cannot expect one nearly that often.

It is often argued that the deaths from air pollution are not very alarming because they are not detectable, and we cannot associate any particular deaths with coal burning. But the same is true of the vast majority of deaths from nuclear reactor accidents. They would materialize only as slight increases of the cancer rate in a large population. Even in the worst accident considered in the Rasmussen Study, expected only once in 100,000 meltdowns, the 45,000 cancer deaths would occur among a population of about 10 million, with each individual's risk being increased by 0.5%. Typically, this would increase his risk of dying from cancer from 20.5% to 21.0%. This risk varies much more than that from state to state—17.5% in CO and NM, 19% in KY, TN, and TX, 22% in NY, and 24% in CT and RI—and these variations are

---

* Note the discussion above indicating that this is based on the highly pessimistic assumption that there will be a failure (or bypass) of the containment in one out of three meltdowns.

rarely noticed. It is thus reasonable to assume that the additional cancer risks even to those involved in this most serious meltdown accident considered in the Rasmussen Study would never be noticed.

If we are interested in *detectable* deaths that can be attributed to an accident, we must limit our consideration to acute radiation sickness which can be induced by very high radiation doses, about a half million millirem in one day or less. It results in death within a month or two. This is a rather rare disease: there were four deaths due to it in the early years among workers in U.S. nuclear programs, but there have been none for the past 15 years.

According to the Rasmussen Study, there would be *no* detectable deaths in 98 out of 100 meltdowns, there would be over 100 such deaths in one out of 500 meltdowns, over 1000 in one out of 5000 meltdowns, and in one out of 100,000 meltdowns there would be about 3500 detectable fatalities.*

The largest number of detectable fatalities to date from an energy-related incident was an air pollution episode in London in 1952 in which 3500 deaths directly attributable to the pollution occurred within a few days.[26] Thus, with regard to detectable fatalities, the equivalent of the worst nuclear accident considered in the Rasmussen Study—expected once in 100,000 meltdowns— has already occurred with coal burning.

But the nuclear accidents we have been discussing are hypothetical, and if we want to consider hypothetical accidents, very high consequences are not difficult to find. For example there are at least two hydroelectric dams in the United States whose sudden rupture would kill over 200,000 people.[7] There are hypothetical explosions of liquefied natural gas that can wipe out a whole city. If we get into possibilities of incubating or spreading germs, or of subtle chemical effects, we can easily imagine even more devastating scenarios.

It is sometimes said that nuclear accidents may be extremely rare, but when they occur they are so devastating as to make the whole technology unacceptable. From the above comparisons it is clear that this argument "holds no water." For another perspective, we are embracing a technology that kills 50,000 Americans *every year,* every one of them clearly detectable, and seriously injuring more than ten times that many—I refer here to motor vehicles. Even if we had a meltdown every ten years, a nuclear power accident would kill that many only once in a million years.

---

* According to some recent analyses,[25] no detectable deaths are expected from any of the accidents considered in the Rasmussen Study.

THE WORST POSSIBLE ACCIDENT

One subject we have not discussed here is the "worst possible nuclear accident," because there is no such thing. In any field of endeavor, it is easy to concoct a possible accident scenario that is worse than anything that has been previously proposed, although it will be of lower probability. One can imagine a gasoline spill causing a fire that would wipe out a whole city, killing most of its inhabitants. It might require a lot of improbable circumstances combining together, like water lines being frozen to prevent effective fire fighting, a traffic jam aggravated by street construction or traffic accidents limiting access to fire fighters, some substandard gas lines which the heat from the fire caused to leak, a high wind frequently shifting to spread the fire in all directions, a strong atmospheric temperature inversion after the whole city has become engulfed in flame to keep the smoke close to the ground, a lot of bridges and tunnels closed for various reasons, eliminating escape routes, some errors in advising the public, and so forth. Each of these situations is improbable, so a combination of many of them occurring in sequence is highly improbable, but it is certainly not impossible. If anyone thinks that is the worst possible consequence of a gasoline spill, consider the possibility of the fire being spread by glowing embers to other cities which were left without protection because their firefighters were off assisting the first city; or of a disease epidemic spawned by unsanitary conditions left by the conflagration spreading over the country; or of communications foul-ups and misunderstandings caused by the fire leading to an exchange of nuclear weapon strikes. There is virtually no limit to the damage that is *possible* from a gasoline spill. But as the damage envisioned increases, the number of improbable circumstances required increases, so the probability for the eventuality becomes smaller and smaller. There is no such thing as the "worst possible accident," and any consideration of what terrible accidents are possible without simultaneously considering their low probability is a ridiculous exercise that can lead to completely deceptive conclusions.

The same reasoning applies to nuclear accidents—accidents causing any number of deaths are possible, but the larger the consequences, the lower is the probability. The worst accident *considered by the Rasmussen Study* would cause about 50,000 deaths, with a probability of one occurrence in a billion years of reactor operation. A person's risk of being a victim of such an accident is 20,000 times less than his risk of being killed by lightning, and a thousand times less than his risk of death from an airplane crashing into his house;[7] no one in his right mind worries about such improbable events.

But this once-in-a-billion-year accident is practically the only nuclear reactor accident ever discussed in the media. When it is discussed, its probability is hardly ever mentioned, and many people, including Helen Caldicott who wrote a book on the subject, say that it's the consequences of an *average* meltdown rather than of one out of 100,000 meltdowns. I have frequently been told that the probability doesn't matter—the very fact that such an accident is *possible* makes nuclear power unacceptable. According to that way of thinking, we have shown that use of gasoline is not acceptable, and almost any human activity can similarly be shown to be unacceptable. If probability didn't matter, we would all die tomorrow from any one of thousands of dangers we live with constantly. But it apparently is only with the risks of nuclear power that probability doesn't matter.

An especially flagrant example of this phenomenon occurred in connection with the November 1982 elections.[27] Congressman Markey, an antinuclear politician from Massachusetts, got a copy of a printout of results from an NRC-sponsored study listing the number of deaths from various nuclear accident scenarios. The worst scenario treated was very extreme, involving highly improbable combinations of situations compounded by failure to take even the most obvious protective actions, but the number of deaths was given for each nuclear power plant in the United States. No probabilities were given or even estimated because the study cut off at probabilities below one in 100 million per year and the probability for those accidents was far below that limit. Markey arranged for these results to be published in the *Washington Post*,[28] from which it was picked up and repeated in just about every newspaper and TV news program in the United States. Markey misinterpreted these data to mean that there was a 2% chance for an accident killing 100,000 people to occur in the United States during this century. The NRC scientists responded that the probability was not more than 0.00002%, 10,000 times smaller than given in the newspaper story, and asked for a correction. The *Washington Post* carried a follow-up story the next day, but retracted nothing and gave no indication of that gross error.

LAND CONTAMINATION

Another aspect of a reactor meltdown accident that has been widely publicized is land contamination. The most common media version is that it would contaminate an area the size of the State of Pennsylvania, 45,000 square miles. Of course this depends on one's definition of "contaminate." It could be said that the whole world is contaminated, because there is natural radioactivity everywhere; or that the State of Colorado is contaminated because the natural radiation there is twice as high as in most other states. However, the Federal Radiation Council in the United States and similar official agencies in other countries have adopted criteria for the level of contamination that is acceptable before people must be evacuated. It corresponds roughly to doubling or tripling the average lifetime dose that would be received from natural radiation and medical X-rays, or 2 to 5 times as much *extra* radiation as would be received by the average American from moving to Colorado. It is still 4 to 10 times less than the natural radiation received by people living in some areas of India and Brazil—studies of these people have given no evidence of health problems from their radiation exposure.[29]

With this definition, the worst meltdown accident considered in the Rasmussen Study—about 1% of all meltdowns might be this bad—would contaminate an area of 3000 square miles, the area of a circle with a 30-mile radius. About 90% of this area could be cleaned up by simply using fire hoses on built-up areas, and plowing the open ground, but people would probably have to be relocated from the remaining 10%, an area equal to that of a circle with a 10-mile radius.[7]

In assessing the impacts of this land contamination, I believe the appropriate measure is the monetary *cost,* the cost of decontaminating, relocating people, compensating for lost property and lost working time, buying up and destroying contaminated farm products, etc. Some might argue that it is unfair to concentrate on money and ignore the human problems in relocation, but that is part of reality. Forced relocation is a common practice in building hydroelectric dams (which flood large land areas), highway construction, slum clearance projects, and so forth, and in these contexts the monetary cost and advantages to be gained are always the prime consideration in deciding on whether to undertake the project. I know people who were forced to relocate for such purposes, and I have not noted any strong feeling that they should have been compensated for their human problems.

In most meltdowns, the cost would be less than $50 million (all costs are in 1975 dollars); in one out of 10 meltdowns, it would exceed $300 million; in one out of 100 meltdowns, it would exceed $2 billion; and once in 10,000 meltdowns, it would be as much as $15 billion.[7]

Over all cases, the average cost would be about $100 million. Production of electricity by coal burning is estimated[30] to do about $600 million per year* in property damage, destroying clothing, eroding building materials, and so forth. Thus it would require six meltdowns per year—one every two

---

* Reference 30 gives $200 million to $2 billion. We use the geometric mean.

months—for the monetary cost to the public from reactor accidents to equal that from coal burning. Clearly, health impacts are more important than property damage in determining the risks of generating electricity, but the relative risks of nuclear power and coal are not very different for the two.

WHY THE PUBLIC MISUNDERSTANDING?

In this chapter we have shown that there is a gross misunderstanding of reactor meltdown accidents in the public mind. In most such accidents there would be *no* harm to the public, and the average meltdown would cause only 400 fatalities and do $100 million in off-site damage. Even in the worst 0.001% of accidents, the increased cancer risk to those involved is much less than that of moving from other parts of the country to New England. This is a far cry from the public image of many thousands of dead bodies lying around in a vast area of devastation, and it certainly is not "the ultimate disaster." Surely only a tiny fraction of the public recognizes that for nuclear accidents to be as dangerous as coal burning, we would have to experience a meltdown every two weeks.

How did this gross public misunderstanding come about? The public gets its information from the media, so the media must be responsible. But why have the media given such a distorted view?

For the first time in the history of our civilization, an industry has exerted an intense effort to determine its environmental effects *in advance*. Thousands of man-years of effort have been expended in dreaming up all the things that could go wrong in nuclear plants and in estimating the possible consequences. With all of this effort, some very damaging scenarios can be developed, although they have very small probabilities.

A media reporter's job is to get an interesting story, something that will attract the attention of the public. If public interest in a nuclear reactor meltdown had not been aroused by the Three Mile Island accident, the facts recited in this chapter would have been dull. A reporter trying to make an interesting story out of it naturally picks the worst accident that has been considered to discuss—45,000 cancers, $15 billion cost. His story may say that a nuclear accident could have these consequences, but readers and those reporting a story seldom make a sharp distinction between *could* and *would*. Since media people get most of their information from other media productions, powerful media figures ranging from Dan Rather to Johnny Carson seem to be convinced that a reactor meltdown is the ultimate disaster. With them preaching this gospel, the public is soon convinced.

Unfortunately, the consequences are tragic. Surely no one believes that we will have a meltdown every two weeks, or even every six months. We have not yet had a single meltdown, and there has never even been a mass evacuation,* although that would be the first action if there appeared to be even a reasonable chance of a meltdown.

Nonetheless, as a result of the fear spread by the media, the U.S. is abandoning nuclear power. We are thus being deprived of our cheapest and safest form of energy at a time when we desperately need it. Instead, utilities are building coal-fired rather than nuclear plants. Every time this is done, many hundreds of Americans are condemned to premature death, not to mention that hundreds of millions of dollars in extra property damage will be incurred.

NON-SAFETY ISSUES

Any new technology is bound to encounter numerous technical problems that must be ironed out, and there has never been any reason to believe that nuclear technology should be an exception in this regard. However, unlike the situation in other industries, technical problems in the nuclear industry often receive widespread media exposure. The media audience has no interest in technical details, so the stories are often spiced up to make the problems seem to be safety issues.

Nearly any technical problem can become a safety issue if it is consistently ignored. If an automobile runs out of lubricating oil, it can stall on a railroad crossing, which is clearly a safety problem. But the oil level is easily checked, there is a warning light indicating loss of oil pressure, and if the oil did run out there would be ominous grinding noises before the car would stall. Moreover, in the great majority of cases there would be no danger if stalling occurred. Loss of lube oil is therefore not ordinarily considered to be a safety problem. It can be inconvenient, costly to fix, and may cause expensive damage to the engine but it surely ranks far down on any list of safety hazards in automobiles. However, if the problem were not so familiar to a large segment of their audience, a media story could easily scare people with stories about the possibility of automobiles stalling on railroad crossings or in other precarious situations due to loss of lube oil.

---

* Mass evacuations following accidents involving dangerous chemicals are relatively common. A few months after the Three Mile Island accident there was such an evacuation in a suburb of Toronto involving over a hundred thousand people for several days.

Analogous situations have happened frequently with technical problems in the nuclear industry. We here review a few of these situations.

Pressurized Thermal Shock[31,32,33]

The thick steel vessel housing the reactor is normally very hot because of the high temperature of the water inside (up to 600°F). If, due to some malfunction, the inside is suddenly filled with cool water, the vessel experiences what is called "thermal shock." If it is then subjected to high pressure—this is pressurized thermal shock (PTS)—there is an increased tendency for it to crack rather than simply to stretch if there is already a small crack or imperfection. The importance of PTS problems, like that of so many others, depends on quantitative details—how much of a thermal shock followed by how much pressure causes how much of an increased tendency to crack? Under ordinary conditions these quantitative details indicate that there is nothing to be concerned about. However, just as radiation can damage biological tissue, it can damage steel by knocking electrons and atoms out of their normal locations. This radiation damage to the reactor vessel aggravates its susceptibility to PTS.

This problem has been recognized for at least 20 years and a simple remedy was found—reducing the quantity of copper in the steel alloy from which the vessel is fabricated. This remedy was implemented in 1971, and all reactor vessels fabricated since that time have no problems with PTS. Reactor vessels are kept under periodic observation to keep track of the problem. For many years, the Nuclear Regulatory Commission (NRC), burdened by other more urgent problems, put off considering PTS by adopting a very conservative screening criterion to indicate when further action on it would be undertaken. In 1981, time for action according to that criterion was only one or two years away in some plants; hence, the NRC began to look into the problem in more detail by requesting information from various power plants. Misinterpreting these requests, the *New York Times* ran a page-one story[34] headlined "Steel Turned Brittle by Radiation Called a Peril at 13 Nuclear Plants," broadly implying that serious safety problems were immediately in prospect. Antinuclear activists soon began trumpeting that message. They claimed that reactor vessels would crumble like glass under pressurized thermal shock, although no such behavior has ever been observed in the numerous laboratory tests of the phenomenon. In 1981–1982, NRC and the nuclear industry delved into the PTS problem rather deeply. In late 1982, NRC came up with new conclusions and proposed regulations.[31]

The condition under which NRC considers the deterioration to be unacceptable is where the threat of a meltdown due to pressurized thermal shock reaches 10% of the already very small threat of a meltdown due to other causes. This 10% estimate contains substantial conservatism; the industry studies[33] suggest that the risk is much smaller. It should be clearly understood that the risk of a meltdown is *not* 10%; rather, the already existing very small risk of a meltdown is increased by no more than 10%.

While no reactor has yet reached this condition, several are expected to approach it during the late 1980s. Three years before they reach it, a plan for remedial action must be filed with NRC. Several remedies are available, although not all are applicable in all situations. One way to postpone the problem is to redistribute the fuel in the reactor so as to reduce the radiation striking the walls of the vessel—this is now being done in several plants. One remedy for PTS is to keep the water storage tanks heated to reduce the thermal shock that would be caused by sudden water injection—this is now being done in one plant. Another remedy is to change operating procedures to reduce the suddenness with which this water can be introduced. The most complete remedy, which is also the most time consuming and expensive, is to heat the reactor vessel to very high temperature (850°F) to anneal out the radiation damage; this would make the vessel as good as new.

There are several elements of conservatism in the formulation of the NRC standard. It is based on the assumption that there is a small crack or flaw in the vessel, although these vessels are very carefully inspected and no small cracks or flaws have been found. The vessel is typically 8 inches thick so the outside is exposed to considerably less radiation and thermal shock than the inside; therefore even if there should be cracking inside, it would probably not extend all the way through the thickness of the vessel and there would consequently be no danger from it.

As long as the problem is recognized, is under constant surveillance, remedies are available, and the situation will not be allowed to reach the danger point, it seems fair to classify pressurized thermal shock as a technical problem rather than as a safety issue. It should therefore receive the attention of scientists and engineers, but there is no actual reason for the public to become involved.

Stress Corrosion Cracking of Pipes

There have been a number of situations in which pipes in boiling water reactors have been found to have cracks.[34] Since a pipe cracking open is a widely heralded potential cause for a LOCA, this received extensive media coverage as a safety problem, especially when the first such crack was dis-covered in 1975. However, it has been established that this type of cracking develops very slowly and is hence usually detected by ultrasonic tests in its very initial stages. If not, it leads to slow leaks which are readily detected and repaired. Stress corrosion cracking is therefore *not* a safety issue.

On the other hand, it has caused expensive shutdowns for repairs, and has therefore been an important problem for power plant owners. They have consequently invested tens of millions of dollars on research to overcome it. The first fruit of this research was to gain an understanding of the problem: welding stainless steel pipe joints was causing some of the chromium that makes that material corrosion resistant to migrate away,* reducing its local concentration from the normal 17% to below the 12% minimum for resistance to corrosion by excess oxygen in the water; moreover, once this migration of chromium is started by the welding, it is continued by the heat of the reactor water. A combination of this corrosion with tensile stress was found to cause the cracking.

Once the problem was understood, solutions were rapidly forthcoming. A new alloy with less carbon and more nitrogen, called nuclear-grade stainless steel, was developed which virtually eliminates the problem in new pipe. It was found that in the old type pipe, the chromium migration could be reversed by heating the welded joint in a furnace to 1950°F, or by putting a lining of weld metal inside the pipe before the outside is welded. In addition to avoiding the chromium migration, methods have been developed to relieve the tensile stress by running cooling water inside the pipe while the joint is being welded, or heating the outside of the pipe while cooling the inside after the welding is completed—this last method is applicable without removing installed pipes. All of these methods are now being applied in operating plants. Moreover, methods are being developed for reducing the free oxygen content in the water, the principal chemical agent responsible for the corrosion. All three elements, chromium migration, mechanical stress, and a corrosive chemical agent, are necessary to cause the cracking, and all three of them have been reduced by these measures. An automated computer-controlled ultrasonic testing system, called "adaptive learning network," has been developed to predict which welds are most likely to fail and estimate their remaining service life. As a result of all this progress, stress corrosion cracking of pipes seems to be well under control.

---

* Even in solid materials, certain atoms can migrate away from their original site. This sometimes happens to the chromium atoms in stainless steel when it is heated.

Steam Generator Tube Leaks[37]

A diagram of a pressurized water reactor (PWR) is shown in Fig. 12. The water in the reactor is kept under sufficiently high pressure that it does not boil to become steam. Rather it is pumped through the tubes of "steam generators" where it transfers its heat to the water from a separate "secondary" system, causing the latter to boil into steam. This has some advantage (and some disadvantages) over the simpler system of generating the steam by boiling the water in the reactor as in the boiling water reactor (BWR). One of the advantages is that the water from the reactor which contains radioactive contaminants never gets into the other areas of the plant (turbine, condenser, etc.) so less attention to radioactivity control is needed in those areas.

However, leaks in steam generator tubes do allow radioactivity to reach those areas, and since they have minimal radioactivity control, it can easily escape from there into the environment. Of the 48 operating PWRs in the United States, 40 have experienced problems with steam generator tube leaks.[37] There are many thousands of these tubes in a steam generator; therefore leaking tubes can simply be plugged-up at both ends without affecting operation. However, when the number of plugged tubes exceeds about 20% of the total, as it has in five plants, the electrical generating capacity of the plant is significantly reduced. This represents a costly loss of revenue to the utility. In at least three cases, the utility has decided to completely replace their steam generators, a rather expensive alternative requiring many months of shutdown.

From the safety viewpoint, the worst accident worthy of consideration in this area is a sudden complete rupture of a few tubes. Such an accident might be expected once every several years. This is what hapopened at the Ginna plant near Rochester, New York in January, 1982. There was a great deal of media publicity for that accident, but the maximum exposure at any offsite point was 0.5 millirem,[36] about what the average American receives from natural sources every two days. Since there were no people at such points, no member of the public received even that much exposure. The total of the exposures to the whole population in the area was less than 100 millirem, which gives only once chance in 80,000 that there will ever be a single cancer resulting. Since that accident represents something approaching the worst from a steam generator tube failure, such accidents can hardly be considered an important threat to public safety.

On the other hand, it has been a costly problem for utilities, and a great deal of research has been devoted to solving it.[37] Eight separate classes of failures have been identified—denting, erosion–corrosion, fatigue, fretting, intergranular attack, pitting, stress corrosion cracking, and wastage. A number of different methods for reducing these problems and for avoiding them in new plants have been developed, as have new methods for detecting, locating, evaluating, and repairing leaks.

The Nuclear Regulatory Commission keeps a close watch on the problem to be certain that public safety is not compromised, in spite of the very small potential of steam generator leaks to cause radiation exposure to the public. It requires frequent testing for leaks, and has strict limits on the amount of leakage that can be tolerated before the reactor is shut down for repairs. It also maintains research programs for improving understanding, evaluations, and predictability of future problems. The industry, of course, is doing much more research on all aspects of the problem.

APPENDIX

Fault Tree Analysis

One question I am often asked is—Since there has never been a reactor meltdown accident, how can you estimate the probability for one?

The method,[7] known as "fault tree analysis," is to identify all "routes" leading to a meltdown, with each route consisting of a succession of failures like pipes cracking, pumps breaking down, valves sticking, operators pushing the wrong button, etc. Since a given route will not lead to meltdown unless *each* of these failures occurs in turn, the probability of meltdown by that route is obtained by multiplying the probabilities for each individual failure. For example, if one particular route to meltdown consists of a pipe cracking badly—expected once in 1000 years of operation—followed by a pump failing to operate—expected once in 100 trials—followed by a valve sticking closed—expected once in 200 attempts to open it, the chance that each of these three failures will occur successively in a given year is

$$\frac{1}{1000} \times \frac{1}{100} \times \frac{1}{200} = \frac{1}{20,000,000}$$

There has been extensive experience in many industries with pipes cracking, pumps failing, and valves sticking, so the probabilities for these are known (The probabilities depend, of course, on the *quality* of the pipes, pumps, and valves, but there is also experience on that). It is these probabilities, obtained from experience, that are used in the calculations.

Once the probabilities for each route to meltdown are calculated, the probabilities for all possible routes must be added up to obtain the total probability for a meltdown. This is the largest source of uncertainty since there is no way to be certain that all possible routes have been included. However, with many dozens of independent researchers thinking about these questions for many years, it seems reasonable to believe that at least most of the important routes have been considered.

The Emergency Core Cooling System—Preventing Meltdown[11]

Following a reactor shutdown, as in an accident, radioactivity in the fuel continues to generate heat at a rate shown by the curve[6] in Fig. 4. Ten seconds after shutdown it is 5% of the full power rate, and this drops to 3% after 3 minutes, 1% after 3 hours, and 0.3% after 5 days. If this heat is not carried away, the fuel will melt.

Under normal circumstances the reactor fuel is submerged in rapidly flowing water which picks up this heat and carries it to some other part of the plant where it is transferred to other systems. There are several things that can go wrong with this routine:

(a) The other system may fail and be unable to accept this heat. (Such a failure, the breakdown of a pump in combination with valves on back-up pumps being left closed for reasons still not explained, initiated the Three

Fig. 13. The rate at which heat is evolved by the radioactivity in the fuel, as percent of the heat evolution rate when the reactor is operating normally, versus time after the accident. The horizontal lines labeled "HPIS" show the amount of heat for which the water it boils off is equal to that supplied by high-pressure injection systems.

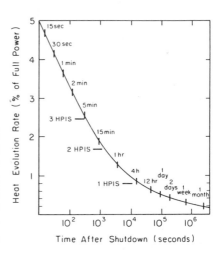

Mile Island accident, although that problem was quickly corrected and would have caused no trouble if it were not for other problems.)

(b) The water flow may be blocked (as by a hydrogen bubble in the late stages of the Three Mile Island accident).

(c) The water may escape slowly through a small leak or open valve. (This was the most important failure in the Three Mile Island accident; a valve failed to close and the operators did not recognize that fact.)

(d) The system may burst open, releasing the pressure and thereby converting the water (which is at 600°F) into steam which would come shooting out through the opening—this is called "blowdown"—leaving the reactor fuel with no water cooling.

In situations (a) and (b), the water will overheat and begin to boil, and the resulting steam will be released through pressure relief valves; these situations therefore also result in loss of water.

If the fuel is not covered with water, it will overheat and eventually melt; thus in all of these situations, it is important to inject more water into the reactor to replace that which is lost. In situations (a), (b), and (c), the reactor remains at high pressure (more than a hundred times normal atmospheric pressure); it therefore requires special high-pressure pumps to inject water into it. There are typically 3 or 4 of these "high-pressure injection systems" (HPIS).[4,7] They provide enough water to make up for boiloff caused by the amounts of heat shown by the horizontal lines labelled "HPIS" in Fig. 12. As an illustration of their meaning, if three HPIS are working, this is enough to match the heat evolving from the fuel after six minutes; after that time, therefore, the total water in the system begins to increase. Up to that time, a calculation shows, only 2% of the water has been boiled away.

If only one of the HPIS is providing water, the rate at which water is provided is not equal to the rate at which it boils off until after 5.5 hours according to Fig. 12. By that time, nearly 40% of the water would be boiled away, but there would still be more than enough left to keep the fuel covered. Thus any *one* of the 3 or 4 HPIS would normally provide enough water to prevent a meltdown, or even damage to the fuel due to overheating. (In the Three Mile Island accident, the HPIS were turned off by the operators because they misinterpreted the information available to them as indicating that there was *too much* water in the reactor.)

In situation (d), the water is lost by blowdown in a matter of seconds; it is therefore important to get a lot of water back into the system immediately. This would be accomplished[4] by 3 or 4 systems called "accumulators." They are large tanks filled with water at about one third the pressure in the reactor. The water in these tanks is normally kept out of the reactor by a valve which is held shut by the higher pressure from the reactor side. But if the pressure in the reactor should fall below that in the accumulator tanks, the latter would push the valve open, rapidly dumping the water from those tanks into the reactor. Note that this is a failproof system, not requiring electric power or any human action. There is enough water in two of the accumulators to keep the fuel covered for about 15 minutes before boil-off would lower its level below the top of the fuel.

Another element of the emergency core cooling system for adding water following a blowdown is two low-pressure injection systems (LPIS), either of which would provide enough water to cover the fuel in about three minutes, and enough to far more than compensate water boil-off at all times.[4,7] If either of these goes into operation by the time the water from the accumulators becomes insufficient, there can be no danger of damage to the fuel due to lack of water. But even if they fail, any one of the LPIS could provide sufficient water for this purpose (they inject much more water when the reactor is at low pressure).

Note that both the HPIS and the LPIS require electric power to drive the pumps. If there should be an electric power failure following a blowdown, there would thus be about 15 minutes—the time during which water from the accumulators keeps the fuel covered—to restore electric power as by starting up one of the diesel generators.

In summary, if even one of the 3 or 4 HPIS works, it is very difficult to imagine a situation in which insufficient water would be provided to prevent damage to the fuel. In the event of blowdown, either one of the LPIS *or* one of the HPIS would avert a meltdown.

Protections against Containment Rupture

In nearly all scenarios for serious reactor accidents, a great deal of water ends up in the containment building. When heat is added to water, the latter can be converted into steam which increases the pressure inside. If this pressure exceeds the maximum that the containment can withstand, the containment will rupture,[11] allowing the release of radioactive material into the environment. The integrity of the containment thus depends on keeping the net heat evolved within the containment below the maximum allowable quantity.

This allowable amount in the absence of containment cooling is shown in Fig. 14 by the curve labeled "no cooling"; it increases with time because some heat is diffusing into the concrete walls where it does not contribute to increasing the steam pressure. The sources of heat are shown by the dashed

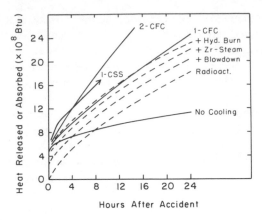

Fig. 14. Total quantity of heat released into the containment (dashed lines) and allowable without rupturing the containment due to high pressure (solid lines), versus time after the accident. The various lines are explained in the text.

lines in Fig. 13. The one source present in all cases is that generated by the radioactivity in the fuel. Its rate of evolution was shown in Fig. 12, and from that curve it is straightforward to calculate the total heat evolved up to any time; that is the dashed curve labeled "radioact" in Fig. 13. If blowdown occurs, the energy from it adds to the total, giving the dashed curve labeled " + blowdown." If the emergency core cooling system fails to restore cooling, there will be a chemical reaction between the fuel cladding material, zirconium, and steam which releases additional heat, bringing the total up to that shown by " + Zr–steam." That reaction generates hydrogen, and if this hydrogen burns, it contributes additional energy, bringing the total to the dashed curve labeled " + Hyd.Burn."

From Fig. 13, we see that the total energy released exceeds the allowable quantity with no cooling after 8 hours even if only the radioactivity contributes, after 4 hours if there is a blowdown, and much sooner if the fuel overheats enough to allow the Zr–steam reaction. Clearly, it is necessary to have systems for cooling the containment atmosphere.

There are two types of systems for doing this.[4] One of these is the containment spray systems (CSS) which spray water into the air to condense the steam. There are typically three of these. They would exhaust their stored water supply after 1–2 hours, leaving the water on the containment floor. The system would then be switched to pick up water from the containment floor, pass it through pipes surrounded by cool water from a separate source outside the containment, and then spray it into the containment air to achieve cooling. If one of these three systems is operating, there can be no danger of breaking the containment due to excess pressure inside. This can be seen from the curve labeled "1-CSS" in Fig. 13 which shows that the heat removed exceeds that provided by all sources combined by a large margin.

Another provision for containment cooling in some power plants is "containment fan coolers" (CFC) which blow air from the containment atmosphere over pipes carrying cooling water from a separate source outside the containment. There are typically five CFC, and the amount to which one or two of these increases the allowable amount of heat input is shown by the curves labeled "1-CFC" and "2-CFC" in Fig. 13. We see that if only one CFC is working, the cooling appears to be slightly insufficient to prevent containment rupture for the first 14 hours if all of the heat inputs contribute fully.* However, if two of the five CFC are working, there is no problem with containment failure due to excess pressure.

REFERENCE NOTES

1. "Report of the President's Commission on The Accident at Three Mile Island," J.B. Kemeny (Chairman), Washington, D.C., October (1979).
2. M. Rogovin (Director), "Three Mile Island, A Report to the Commissioners and to the Public," Washington, D.C., January (1980).
3. J. R. Lamarsh, *Introduction to Nuclear Engineering* (Addison-Wesley, Reading, Massachusetts, 1975); S. Glasstone and W. H. Jordan, *Nuclear Power and its Environmental Effects* (American Nuclear Society, La Grange Park, Ill., 1980).
4. G. Masche, "Systems Summary of a Westinghouse Pressurized Water Reactor Nuclear Power Plant," Westinghouse Electric Co. (1971).
5. S. Hoffman and T. Moore, "'General Description of a Boiling Water Reactor," General Electric Co. (1976).
6. "American National Standard for Decay Heat Power in Light Water Reactors," American National Standards Inst. ANSI/ANS-5.1-1979.
7. "Reactor Safety Study," Nuclear Regulatory Commission Document WASH-1400, NUREG 75/014 (1975).
8. "Analysis of Three Mile Island—Unit 2 Accident," Nuclear Safety Analysis Center Report NSAC-1, Palo Alto, California, July (1979); "Nuclear Accident

---

* The curves in Fig. 13 are calculated with conservative assumptions about the cooling capacity of the CFC (they are based on 56 million BTU/hr whereas the units are usually designed for 80 milion BTU/hr heat removal) and the containment rupture pressure (they are based on 100 psi whereas best estimates are 150 psi). One CFC would therefore very probably be sufficient in all cases.

and Recovery at Three Mile Island," Senate Committee on Environment and Public Works, Serial No. 96-14, July (1980); "Investigation of the March 28, 1979 Three Mile Island Accident," U.S. Nuclear Regulatory Commission Document NUREG-0600 (Aug., 1979); "Three Mile Island: The Most Studied Nuclear Accident in History," Report to the Congress by the Comptroller-General, U.S. General Accounting Office Report EMD-80-109, September 9 (1980).

9. Report of the Special Review Group, "Lessons Learned from Three Mile Island," U.S. Nuclear Regulatory Commission Document NUREG-0616, December (1979).

10. "The Safety of Nuclear Power Plants and Related Facilities," U.S. AEC Report WASH-1250, July (1973).

11. B. L. Cohen, "Physics of the Reactor Melt-down Accident," *Nuclear Science and Engineering,* **80,** 47 (1982).

12. R. Gillette, "Nuclear Reactor Safety," *Science,* **176,** 492, 5 May (1972); **177,** 771, 1 Sept.; **177,** 867, 8 Sept.; **177,** 970, 15 Sept.; **177,** 1080, 22 Sept.

13. I. Forbes, J. MacKenzie, D. F. Ford, and H. W. Kendall, "Cooling Water," *Environment,* January (1972), p. 40; D. F. Ford, and H. W. Kendall, "Nuclear Safety," *Environment,* September (1972).

14. M. L. Russel, C. W. Solbrig, and G. D. McPherson, "LOFT Contribution to Nuclear Power Reactor Safety and PWR Fuel Behavior," *Proceedings of the American Power Conference,* **41,** 196 (1979); J. C. Lin, "Post Test-Analysis of LOFT Loss of Coolant Experiment L2-3," EG&G Idaho Report EGG-LOFT-5075 (1980); J. P. Adams, "Quick Look Report on LOFT Nuclear Experiment L2-5," EG&G Idaho Report EGG-LOFT-5921 (1982).

15. S. McCracken, *The War Against the Atom* (Basic Books , New York, 1982).

16. W. Marshall (Chairman of Study Group), "An Assessment of the Integrity of PWR Pressure Vessels," U.K. Atomic Energy Authority, March (1982).

17. W. A. Carbiener *et al.,* "Physical Processes in Reactor Melt-down Accidents," Appendix VIII to Nuclear Regulatory Commission Document WASH-1400 (1975).

18. W. R. Butler, C. G. Tinkler, and L. S. Rubinstein, "Regulatory Perspective on Hydrogen Control for LWR Plants," Workshop on Impact of Hydrogen on Water Reactor Safety, Albuquerque, New Mexico, January (1981); W. R. Butler and C. G. Tinkler, "Regulatory Perspective on Hydrogen Control for Degraded Core Accidents," Second International Workshop on the Impact of Hydrogen on Water Reactor Safety, Albuquerque, New Mexico (1982).

19. "Hydrogen Control for Sequoyah Nuclear Plant," Nuclear Regulatory Commission Document dated August 13 (1980).

20. "Proposed Interim Hydrogen Control Requirements for Small Containments," Memorandum from H. Denton to The NRC Commissioners dated February 22 (1980), NRC Document SECY-80-107.

21. M. R. Fleishman, W. R. Butler, J. T. Larkins, and M. A. Taylor, "Status of Hydrogen Control Activities," NRC Document dated September 3 (1981).

22. Union of Concerned Scientists, "The Risks of Nuclear Power Reactors," Cambridge, Massachusetts (1977).

23. H. W. Lewis (Chairman), "Risk Assessment Review Group Report to the U.S. Nuclear Regulatory Commission," NUREG/CR-400 (1978).
24. "Mitigation of Small-Break LOCAs in Pressurized Water Reactor Systems," Nuclear Safety Analysis Center (Palo Alto, California) Document NSAC-2 (1980); E. Zebroski, Nuclear Safety Analysis Center, private communication (1981); R. J. Breen, Nuclear Safety Analysis Center, private communication (1983).
25. F. J. Rahn and M. Levenson, "Radioactivity Releases Following Class-9 Reactor Accidents," Health Physics Society, Las Vegas, Nevada, June (1982); C. D. Wilkinson, "NSAC Workshop on Reactor Accident Iodine Release," Palo Alto, California, July (1980); H. A. Morewitz, "Fission Product and Aerosol Behavior Following Degraded Core Accidents," *Nuclear Technology* 53, 120 (1981).
26. U.S. Senate Committee on Public Works, "Air Quality and Stationary Source Emission Control" (1975); R. Wilson, S. D. Colome, J. D. Spengler, and D. G. Wilson, *Health Effects of Fossil Fuel Burning* (Ballinger Publ. Co., Cambridge, Massachusetts, 1980).
27. Nuclear Report, *American Nuclear Society,* 5(10), November 5 (1982).
28. M. R. Benjamin, "Nuclear Study Raises Estimates of Accident Tolls," *Washington Post,* November 1, (1982).
29. International Symposium on Areas of High Natural Radioactivity, Academy of Sciences of Brazil, June (1975).
30. W. Ramsay, *The Unpaid Costs of Electrical Energy* (Johns Hopkins University Press, Baltimore, Maryland, 1979).
31. "Draft NRC Evaluation of Pressurized Thermal Shock," September 13 (1982).
32. T. A. Meyer, "Summary Report on Reactor Vessel Integrity for Westinghouse Operating Plants," Westinghouse Electric Corp. Report WCAP-10019, December (1981).
33. "Summary of Evaluations Related to Reactor Vessel Integrity Performed for the Westinghouse Owner's Group," Westinghouse Electric Corp., Nuclear Technology Division, May (1982).
34. M. L. Wald, "Steel Turned Brittle by Radiation Called a Peril at 13 Nuclear Plants," September 27 (1981).
35. R. Immel, "Stress Corrosion Cracking," *EPRI Journal,* November (1981).
36. "Report on the January 25, 1982 Steam Generator Tube Rupture at the R.E. Ginna Nuclear Power Plant," NRC Document NUREG-0909, April (1982).
37. "Steam Generator Tube Experience," U.S. Nuclear Regulatory Commission Document NUREG-0886 (1982).

Chapter 4 / UNDERSTANDING RISK

The third major reason for public misunderstanding of nuclear power is that the great majority of people do not understand and quantify the risks we face. Most of us think and act as though life is largely free of risk. We view taking risks as foolhardy, irrational, and assiduously to be avoided. Training children to avoid risk is an all-important duty of parenthood. Risks imposed on us by others are generally considered to be entirely unacceptable.

Unfortunately, life is not like that. Everything we do involves risk.[1] There are dangers in every type of travel, but there are dangers in staying home—40% of all fatal accidents occur there.[2] There are dangers in eating—food is probably the most important cause of cancer and of several other diseases—but most people eat more than necessary. There are dangers in breathing—air pollution probably kills at least 10,000 Americans each year,[3] inhaling natural radioactivity is believed to kill a similar number,[4] and many diseases are contracted by inhaling germs. These dangers can largely be avoided by breathing through filters, but no one does that. There are dangers in working—12,000 Americans are killed each year in job-related accidents, and probably ten times that number die from job-related illness[5]—but most alternatives to working are even more dangerous. There are dangers in exercising and dangers in not getting enough exercise. Risk is an unavoidable part of our everyday lives.

That doesn't mean that we should not try to minimize our risks, but it is important to recognize that minimizing anything must be a *quantitative* procedure. We cannot minimize our risks by simply avoiding those we happen to think about. For example, if one thinks about the risk of driving to a destination, he might decide to walk, which in most cases would be much more dangerous.[2] The problem with such an approach is that the risks we think about are those most publicized by the media, whose coverage is a very poor guide to actual dangers. The logical procedure for minimizing risks is to quantify all risks and then choose those that are smaller in preference to those that are larger. The main object of this chapter is to provide a framework for that process and to apply it to nuclear power risks.

There are many ways of expressing quantified risk, but here we will use just one, the loss of life expectancy (LLE); i.e., the average amount by which one's life is shortened by the risk under consideration. The LLE is the product of the probability for a risk to cause death and the consequences in terms of lost life expectancy if it does cause death. As an example, statistics indicate[6] that an average 40 year old will live another 34.8 years, so if he takes a risk that has a 1% chance of being immediately fatal, it causes an LLE of 0.348 years (0.01 × 34.8).

It should be clear that this does *not* mean that he will die 0.348 years sooner as a result of taking this risk. But if 1000 people his age took this risk, 10 might die immediately, having their lives shortened by 34.8 years, while the other 990 would not have their lives shortened at all. Hence, the *average* lost lifetime for the 1000 people would be 0.348 years. This is the LLE from that risk.

Of course most risks are with us to varying extents at all ages and the effects must be added up over a lifetime, which makes the calculations somewhat complex. We therefore developed a computer program for doing the calculations and used it to carry out a rather extensive study of a wide variety of risks. Some of the results of that study are summarized in the next section.

A CATALOG OF RISKS[1]

One of the greatest risks in our society is remaining unmarried. Statistics[7] show that a single white male has a life expectancy of 62.3 years versus 68.3

years for the married white male,* corresponding to an LLE of 6.0 years. For single white females and nonwhite males and females, the LLEs are 3.2, 8.8, and 6.0 years respectively. Note that males suffer much more than females from remaining single and blacks suffer more than whites. One might suspect that part of the reason for these differences is that sickly people are less likely to marry, but this is evidently not the main reason, since mortality rates are even higher for widowed and divorced people at every age. The life expectancy for a 55-year-old white male is 19.6 years if he is married, 16.4 years if he is single, 15.7 years if he is widowed, and 13.4 years if he is divorced, assuming in the last three cases that he does not marry later. For white females these figures are 24.9, 23.0, 22.2, and 22.4 years, respectively.

A much more widely recognized risk is cigarette smoking.[8] For one pack per day, this has an LLE of 6.4 years for men and 2.3 years for women—in the former case this corresponds to an LLE of 10 minutes for each cigarette smoked. For noninhalers, the lifetime risk from one pack per day is 4.5 years to men and 0.6 years for women, while for those who inhale deeply it is 8.6 years for men and 4.6 years for women. (The differences between male and female risks may involve *how deeply* they inhale, or some of their differences in lifestyle and physiology.) Giving up smoking reduces these risks: after five years the LLE is reduced by one-third, and after ten years it is more than cut in half. Cigar and pipe smoking do little harm if there is no inhalation, but with inhalation the LLE is 1.4 years for pipes and 3.2 years for cigars.

Further understanding of the risks in smoking comes from examining the diseases from which smokers die more frequently.[8] We give figures on the ratio of death rates for 1–2 packs per day smokers to nonsmokers in the 35–84 age range. This ratio is 17 for lung cancer (i.e., heavy smokers are 17 times more likely to die from lung cancer than nonsmokers), 13 for cancer of the pharynx and esophagus, 6 for cancer of the mouth, 11 for bronchitis and emphysema, 4 for stomach ulcer, 3 for cirrhosis of the liver, 2 for influenza and pneumonia, 1.8 for cardiovascular disease, our nation's No. 1 killer, and between 1.5 and 2 for leukemia and cancer of the stomach, pancreas, prostate, and kidney.

Another major risk over which we have some personal control is being overweight[9]—we lose about one month of life expectancy for each pound our weight is above average for our size and build. In the case of someone 30 pounds overweight, the LLE is thus 30 months or $2\frac{1}{2}$ years. To assess the

* These figures are based on 1968 statistics when the study was made on marital status. According to 1979 statistics, the life expectancy for the average white male is 70.6 years.

effect of overeating, we note that our weight increases by 7 pounds for every 100 calorie increase in average daily food intake.[10] That is, if an overweight person changes nothing about his eating and exercise habits except for eating one extra slice of bread and butter (100 calories) each day, he will gain 7 pounds (gradually over a period of about one year) and his life expectancy will be reduced by 7 months. This works out to a 15 minute LLE for each 100 extra calories eaten.

Any discussion of major risks must include the traditional leader, disease.[11] Heart disease leads in this category with LLE-5.8 years, followed in order by cancer with LLE-2.7 years, stroke with LLE-1.1 years for men and 1.7 years for women, pneumonia and influenza with LLE-4½ months, and cirrhosis of the liver and diabetes with LLEs of a little over three months each, the former more in men and the latter more in women by about 3 to 2 ratios.

Perhaps the least appreciated of all major risks is that of being poor, unskilled, and/or uneducated. The best data are on occupational groupings.[12] Professional, technical, administrative, and managerial people live 1.5 years longer than those engaged in clerical, sales, skilled, and semiskilled labor, and the latter group lives 2.4 years longer than unskilled laborers. Corporation executives live 3 years longer than even the longest-lived of the above groups, a full 7 years longer than unskilled laborers. A similar study in England[13] showed even larger differences, finding comparable differences among wives of workers; the wife of a professional person lives about four years longer than the wife of an unskilled laborer. This indicates that the problems are not occupational exposures but rather socioeconomic.

In seeking to understand the reasons for these differences, it is interesting to consider the causes of death. If we compare unskilled laborers with professional, technical, administrative, and managerial people in the United States,[12] their risk of early death from tuberculosis is 4.2 times higher, from accidents is 2.9 times higher, from influenza and pneumonia is 2.8 times higher, from cirrhosis of the liver is 1.8 times higher, and from suicide is 1.7 times higher. It is also 30% higher from cancer and 13% higher from cerebrosvascular disease, but it is 8% lower from arteriosclerosis and diabetes. The large factors in this list are from causes associated with unhealthy living conditions, limited access to medical treatment or unenlightened attitudes toward health care, and are thus generally preventable. This would seem to be a fertile field for social action.

A similar pattern appears in correlations between life expectancy and educational attainment.[14] College-educated people live 2.6 years longer than the average American, while whose who dropped out of grade school live

1.7 years less than average, a 4.3-year differential. These differences are about the same for men and women, which indicates that occupational exposures are not the basic problem here. Dropping out of school at an early age ranks with taking up smoking as one of the most dangerous acts a young person can perform. Even volunteering for combat duty in wartime pales by comparison; the LLE from being sent to Vietnam during the war there[1] was 2.0 years in the Marines, 1.1 year in the Army, 0.5 years in the Navy, and 0.28 years in the Air Force.

Life expectancy varies considerably with geography in ways not explainable by socioeconomic differences. For whites it is over a year longer than average in the rural north central states North Dakota, South Dakota, Minnesota, Iowa, Nebraska, Kansas, and Wisconsin (all of which have lower than the national average income), while it is a year shorter than average in the rural southeastern states South Carolina, Georgia, Alabama, Mississippi, and Louisiana.

The most highly publicized risks are those of being killed in accidents[2]—the suddenness and drama of accidental death are well suited to the functions of our news media—although the actual danger is well below that of the risks we have discussed previously. The LLE from all accidents combined is 435 days (1.2 years). Almost half of them involve motor vehicles which give us an LLE of 207 days, 170 days while riding and 37 days as pedestrians.* Using small cars rather than standard size increases one's LLE by 50 days, and changing from standard size to large cars reduces it by an equal amount.[15] Before the national speed limit was reduced from 65 to 55 miles per hour, the total LLE was 40 days higher. On an average, riding one mile in an automobile and crossing a street each have an LLE of 0.4 minutes, making them as dangerous as one puff on a cigarette (assuming 25 puffs to a cigarette), or, for an overweight person, eating 3 extra calories.

The total LLE over a lifetime from various other types of accidents[2] is 40 days each for falls (mostly among the elderly) and drowning, 27 days for fire and burns, 17 days for poisoning, 13 days for suffocation, 11 days for accidents with guns, and $7\frac{1}{2}$ days for asphyxiation. Men are more than twice as likely to die in accidents as women,[16] in motor vehicle accidents the male/female ratio is $2\frac{1}{2}$ to 1 for both riders and pedestrians, and in drowning the ratio is 5 to 1.

Accidental death rates vary greatly with geography[17]; they are 4 times higher in Wyoming than in New York to give the two extremes; the Northeast

---

* The risk per mile traveled is higher for pedestrians, but we travel far more miles in motor vehicles than as pedestrians.

is generally the safest area while the Rocky Mountain states are generally the most dangerous.

We spend most of our time at home and at work, so that is where most of our accidents occur that are not related to travel. The LLE for accidents in the home is 95 days, and for occupational accidents it is 74 days.[2] The latter number varies considerably from industry to industry, from about 300 days in mining, quarrying, and construction to 30 days in trade, e.g., clerks in stores. Nearly half of all workers are in manufacturing and service industries for which the LLE is 45 days. For radiation workers in the nuclear industry, radiation exposure gives them an average LLE of 12 days.[5]

Actually, these statistics cover up many high-risk occupations because they average over whole industries including white collar workers and many others in relatively safe jobs.[5] Canadian occupational accident statistics[18] are kept in much finer detail, and elucidate some of these effects:

- in the mining industry, the LLE for those who sink shafts is 660 days versus 65 days for those involved in shop work and service;
- in the utility industry, the LLE for those who work with power lines is 820 days versus 58 days for mechanics and fitters;
- in forestry, the LLE for those who fell trees is 1050 days versus 54 days for sawmill workers;
- in construction, the LLE for demolition workers is 1560 days (more than 4 years) versus 38 days for those involved with heating, plumbing, and electrical wiring.

Some showmanship activities are widely advertised as having very high accident potential, but judging from statistical experience, these[19] dangers are exaggerated in the public mind. Professional aerialists—tight-rope walkers, trapeze artists, aerial acrobats, and high-pole balancers—get an LLE of 5 days per year of participation, or 100 days from a 20 year career. The risk is similar for automobile and motorcycle racers of various sorts. The risk of accidental death in these professions therefore is less than in ordinary mining and construction work. The most dangerous profession involving thousands of participants is deep-sea diving, with an LLE of 40 days per year of participation.

In addition to accidents, occupational exposure causes many diseases that affect a worker's life span[5,20] which in most cases are much more important than accidents. Coalminers, on an average, live three years' shorter lives than the average man in the same socioeconomic status, and statistics are similarly unfavorable for truckers, fishermen, ship workers, steel erectors, riggers, actors and musicians (perhaps due to irregular hours), policemen, and firemen. On the other hand, there are occupational groups in which men live a year or more *longer* than average for their socioeconomic standing, such as postal workers, government officials, university teachers, and gardeners. Clearly one's choice of occupation can have a large effect on one's life expectancy, extending to several years.

Homicide and suicide are significant risks in our society, with LLEs of about 135 days each for men, and 43 and 62 days, respectively, for women.[6] Homicide is more common among the young while suicide becomes several times more important among the elderly.

Judging by the media coverage they attract, one might think that large catastrophes pose an important threat to us, but this is hardly the case.[21] Hurricanes and tornadoes combined give the average American an LLE of 1 day, as do airline crashes. Major fires and explosions (those with 8 or more fatalities) give us an LLE of 0.7 days, and our LLE from massive chemical releases is only 0.1 day.

The media have publicized the dangers of various individual substances from time to time. Coffee is believed to cause bladder cancer, with an LLE of 6 days for regular users.[22] There is some evidence that saccharin may cause bladder cancer[23]; the LLE from one diet soft drink every day of one's life is 2 days due to the saccharin, but the weight gain from one extra nondiet soft drink per day causes an LLE of 200 days. Birth control pills can cause phlebitis[22] which gives its users an LLE of 5 days.

Even very tiny risks often receive extensive publicity. Perhaps the best example was the impending fall of our orbiting Sky-Lab satellite, which gave us an LLE of 0.002 seconds.[24] There has been heavy publicity for leaks from radioactive waste burial grounds, although these have not given any single member of the public an LLE as large as 10 seconds.[24] It is shown in the Appendix that the Three Mile Island nuclear power plant accident gave the average Harrisburg area resident an LLE of 1.5 minutes (0.001 days). Our risk of being struck by lightning[21] gives us an LLE of 20 hours.

There are several very large risks that are so mundane that we often ignore them. Females live nearly eight years longer than males and whites live 5.5 years longer than blacks.[6] One might therefore say that the LLE from being a male rather than a female is 8 years, larger than for any other risk we have considered, and the LLE due to being black is 5.5 years although much of this may be due to socioeconomic factors.

For convenient reference, some of the LLEs we have discussed are summarized in Table 1, and shown graphically in Fig. 15.

TABLE 1
LOSS OF LIFE EXPECTANCY (LLE) DUE TO VARIOUS RISKS

Activity or risk	Days LLE
Being male rather than female	2800
Heart disease	2100
Being unmarried	2000
Being black rather than white (in U.S.)	2000
Cigarettes (1 pack/day)	1600
Working as a coalminer	1100
Cancer	980
30 lb overweight	900
Grade-school dropout	800
Being poor	700
Stroke	520
~ 15 lb overweight	450
All accidents	435
Vietnam army duty	400
Living in southeastern U.S. (South Carolina, Mississippi, Georgia, Louisiana, Alabama)	350
Mining or construction work (due to accidents only)	320
~ Motor vehicle accidents	200
Pneumonia, influenza	130
Alcohol	130
Suicide	95
Homicide	90
Occupational accidents (average)	74
Small cars (versus standard size)	50
Drowning	40
~ Speed limit 55 → 65 mph	40
Falls	39
Poison + suffocation + asphyxiation	37
Fire, burns	27
Radiation worker, age 18–65	12
~ Firearms	11
~ Diet drinks (one per day throughout life)	2
All electric power in U.S., nuclear (UCS)	1.5
Hurricanes, tornadoes	1
~ Airline crashes	1
Dam failures	0.5
Spending lifetime near nuclear power plant	0.4
All electric power in U.S. Nuclear (NRC)	0.03

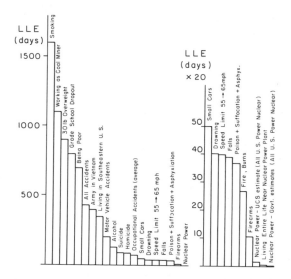

Fig. 15. Graphic representation of relative magnitudes of various risks.

RISKS OF NUCLEAR ENERGY—IN PERSPECTIVE

With the benefit of this perspective, we now turn to the risks of nuclear energy, and evaluate them as if *all* electricity now used in the United States were generated from nuclear power. The calculations are explained in the Appendix, but here we will only quote the results.

According to the study by the U.S. Nuclear Regulatory Commission (NRC) discussed in Chapter 3, the risk of reactor accidents would reduce our life expectancy by 0.012 days or 18 *minutes,* whereas the UCS estimate[25] is 1.5 days. Since our LLE from being killed in accidents is now 435 days, it would be increased by 0.003% according to NRC, or by 0.3% according to UCS. This makes nuclear accidents tens of thousands of times less dangerous than moving from the Northeast to the West (where accident rates are much higher)[17] an action taken in the last few decades by millions of Americans with no consideration given to the added risk. Yet nuclear accidents are what a great many people are worrying about.

The only other comparably large health hazard due to radiation from the nuclear industry is from radioactivity releases into the environment during routine operation. Typical estimates[26] are that, with a full nuclear power program, this might eventually result in average annual exposures of 0.2

mrem (it is now less than one-tenth that large), which would reduce our life expectancy by another 17 minutes. This brings the total LLE from nuclear power to 35 minutes (with this 17 minutes added, the UCS estimate is still about 1.5 days).

If we compare these risks with some of those listed in Table 1, we see that having all nuclear electricity in this country would present the same added health risk (UCS estimates in brackets) as a regular smoker indulging in one extra cigarette every 15 years [every 3 months], or as an overweight person increasing his weight by .012 ounces [0.8 ounces], or as in raising the U.S. highway speed limit from 55 miles per hour to 55.006 [55.4] miles per hour, and it is 2000 times [30 times] less of a danger than switching from standard size to small cars. Note that these figures are *not* controversial because I have given not only the estimates of "establishment" scientists but also those of the leading antinuclear activist group in this country, UCS.

I have been presenting these risk comparisons at every opportunity for several years, but I get the impression that they are interpreted as the *opinion* of a nuclear advocate. Newspapers often say "Dr. Cohen claims . . .". But there is no personal opinion involved here. Deriving these comparisons is simple and straightforward mathematics which no one can question. I have published them in scientific journals, and no scientist has objected to them. I have quoted them in debates with three different UCS leaders and they have never denied them. If anyone has any reason to believe that these comparisons are not valid, they have been awfully quiet about it.

It seems to me that these comparisons are the all-important bottom line in the nuclear debates. Nuclear power is being rejected because it is viewed as being too risky, but the best way for a person to understand a risk is to compare it with other risks with which he is familiar. These comparisons are therefore the best way for members of the public to understand the risks of nuclear power. All of the endless technical facts thrown at them are unimportant and unnecessary if they only understand these few simple risk comparisons. That is all they really need to know about nuclear power. But somehow they are never told these facts. The media never give them, and even nuclear advocates hardly ever quote them. Instead the public is fed a mass of technical and scientific detail that it doesn't understand, which therefore serves to frighten.

When I started may investigations into the safety of nuclear energy in 1971, I had no preconceived notions and no "axes to grind." I was just trying to understand in my own way what the fuss was all about. Rather early in these efforts, I started to develop these risk comparisons. They convinced me that nuclear power is acceptably safe with lots of room to spare. If I am a nuclear advocate, it is because developing these comparisons has made me so.

To be certain that this all-important bottom line is not missed, let me review it. According to the best estimates of "establishment" scientists, having all the electricity now used in the United States generated by nuclear power plants would give the same risk to the average American as a regular smoker indulging in one extra cigarette every 15 years, as an overweight person increasing his weight by 0.012 ounces, or as raising the U.S. highway speed limit from 55 to 55.006 miles per hour, and it is 2000 times less risky than switching from standard size to small cars. If you do not trust "establishment" scientists and prefer to accept the estimates of the Union of Concerned Scientists, the leading antinuclear activist organization in the United States and scientific advisors to Ralph Nader, then having all U.S. electricity nuclear would give the same risk as a regular smoker smoking one extra cigarette every three months, or of an overweight person increasing his weight by 8/10 of an ounce, or of raising the U.S. highway speed limit from 55 to 55.4 miles per hour, and it would still be 30 times less risky than switching from standard size to small cars.

The method for determining these numbers is explained in the Appendix.

ACCEPTABILITY OF NUCLEAR POWER RISKS

The purpose of the discussion presented above is to make the risks of nuclear power *understandable*. Risks are best understood when compared to other risks with which we are familiar. But we have not discussed the question of whether they are *acceptable*. Acceptability includes other factors than the magnitude of risks. People are more willing to accept voluntary risks like skiing, auto racing, and mountain climbing than involuntary risks like pollution,[27] and antinuclear activists are quick to point out that nuclear power presents an involuntary risk to the public. On the other hand, most of the other risks we have discussed are involuntary or at least have an important involuntary component. Poor people certainly are not poor by choice. Living in Southeastern states is no more voluntary than living near a nuclear plant. In many if not most cases, a person's occupation is determined more by circumstances than by voluntary choice; a boy who grows up in the coal fields has often not been exposed to other occupational options and has little opportunity to explore them, so his becoming a coal miner is far from voluntary.

Riding in automobiles is hardly voluntary for most people, as they have no other way to get to work, to purchase food, and to participate in other normal activities of life; even if one avoids riding in automobiles, he is still subject to accidents to pedestrians which account for 20% of deaths from motor vehicle accidents. A large fraction of other accidents are largely due to involuntary activities. Most drownings are of children, but a parent cannot prevent his child from going swimming without risking psychological damage to the child. An appreciable number of drownings result from taking baths, which is hardly a voluntary activity. Deaths from fire, burns, falls, poison, suffocation, and asphyxiation are also not usually due to voluntary risk taking.

Some people are more willing to accept natural risks than man-made risks, but nearly all of the risks we have considered are man-made. Living with artificial risks is the price we pay for the benefits of civilization.

Some say that risks which occur frequently but kill only one or a few people at a time are far less important than occasional large catastrophes which kill the same total number of people. That is the prima facie attitude of the media since their coverage is certainly much greater for the latter than for the former. Based on this viewpoint, some people attribute greater importance to the very rare reactor meltdown accident in which there is large loss of life, than to air pollution which kills far larger numbers of people one at a time.

Actually this argument is highly distorted. The cancers from even the worst meltdown accident considered in the Rasmussen Study would not be any more noticeable than deaths now occurring from air pollution; it would increase the cancer risk of those exposed by only about half of one percent, whereas normal cancer risks are 20%, varying by several percent with geography, race, sex, socioeconomic status, etc. Noticeable fatalities are expected in only two percent of all meltdowns,[21] and even the worst meltdown treated in the Rasmussen Study is expected to cause only a few thousand such deaths. There has already been a comparable disaster from coal burning, an air pollution episode in London in 1952 in which there were 3500 more deaths than normally expected within a few days.[3]

Since reactor meltdowns are *potential* accidents, none having occurred, they should be compared with other *potential* accidents. There are dam failure accidents[21] which could kill 200,000 people within a few hours; they are estimated to be far more probable than a bad nuclear meltdown accident. There are many potential causes of large loss of life anywhere large numbers of people congregate. A collapse of the upper tier of a sports stadium, a fire in a crowded theater, and a poison gas getting into a ventilation system of a large building (some buildings house 50,000 people) are a few examples. Any of these are far more likely than the fraction of one percent of all nuclear meltdowns that would cause large loss of life. The idea that a potential reactor meltdown accident is uniquely or even unusually catastrophic is grossly erroneous.

But I have a deeper objection to the idea that a catastrophic accident is more important than a large number of people dying unnoticeably. To illustrate, suppose there are two technologies, A and B, for producing a desired product, that Technology A causes one large accident per year in which 100 people are killed, and that Technology B kills 1000 people each year one at a time and unnoticeably. If Technology B is chosen to avoid the catastrophic accidents from Technology A, 900 extra people die unnecessarily each year. How would you like to explain to these 900 people and their loved ones that they must die because the media like to publicize large catastrophes? Moreover, any one of us might be one of these 900 that die unnecessarily each year. I therefore maintain that in choosing between technologies on the basis of health impacts, the *total number of deaths* or the LLE should be the overriding consideration.

If we are forced to accept the overblown importance of catastrophic accidents, at the very least it should be clearly recognized that the media are responsible for large numbers of needless deaths.

What risks are acceptable is not a scientific question, and I as a scientist therefore cannot claim expertise on it. I have merely presented the risks as they are, I hope in understandable terms. If any citizen feels that the benefits of electricity produced by nuclear power plants are not worth the risk to the average citizen of a regular smoker smoking one extra cigarette every several months (or years), or of an overweight person adding a fraction of an ounce to his weight, or of raising the national speed limit from 55 to something like 55.1 miles per hour, he is entitled to that opinion. However, he is then obligated to suggest a substitute for the nuclear electricity. Perhaps the next two sections will aid him in this endeavor.

RISKS FROM AIR POLLUTION IN COAL BURNING

It is clearly counterproductive to eliminate one risk if doing so introduces other risks that are greater. In our effort to minimize risk, we must therefore consider the risks of alternatives. In this section we consider the coal-burning alternative, and other alternatives are treated in the next section.

If a utility finds it necessary to build a large new electric power plant these days, it usually has only two options: it can build a nuclear or a coal-burning plant. The important question from the standpoint of public health is, therefore, which of these two alternatives has the greater health risks. We have already discussed the nuclear risk, so now we turn to the risk from coal burning, which, at least for the forseeable future, is largely dominated by air pollution.

The clearest evidence linking air pollution to increased mortality comes from several catastrophic episodes in which a large number of excess deaths occurred during times of high pollution levels.[3] In a December 1930 episode in the Meuse Valley of Belgium, there were 60 excess deaths and 6000 illnesses. In an October 1948 episode in Donora, Pennsylvania, there were 20 deaths (versus 2 normally expected) in a 4-day period during which 6000 of the 14,000 people in the valley became ill. There were at least 8 episodes in London between 1948 and 1962 in each of which hundreds of excess deaths were recorded, the largest in December 1952 when 3500 died. There were three episodes in New York City involving over a hundred deaths, one in November 1953 causing 360 deaths, another in January–February 1963 leading to 500 deaths, and a third in November 1966 responsible for 160 deaths.[28]

The best method for establishing a connection between "normal" levels of air pollution and premature mortality is through comparison of mortality rates between different geographic areas with different air pollution levels. Of course there are other factors affecting mortality rates that vary with geographic area, like socioeconomic conditions; the data must be analyzed thoroughly to eliminate these. As an example, Fig. 16 shows a plot of annual mortality rates for males aged 50–69 in various census tracts of Buffalo, New York during 1959–1961 versus average income and average air pollution (level of Total Suspended Particulate matter—TSP).[29] It is evident from Fig. 16 that for each income range, mortality rates increased with increasing pollution level. A mathematical analysis separates the two effects, giving the risk of air pollution alone. There have been a number of other such studies,[30] comparing various cities in the United States, all the counties in the United States, various cities in England, and so on. In addition there have been a number of studies[31] of mortality rates in a given city, especially New York and London, but also in several other American cities and Tokyo, on a day-by-day basis correlating them with air pollution levels. In these there are no complications from socioeconomic factors, since these do not vary on a day-to-day basis, but there are weather factors that must be removed by mathematical analysis to determine the effects of air pollution alone.

These studies have established strong correlations between air pollution levels and mortality rates. There are also numerous studies[32] of temporary illness, involving hospital admissions, questionnaires, measurements of pulmonary function, etc. in New York, London, Chicago, five Japanese cities,

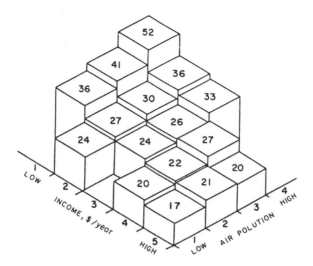

Fig. 16. Mortality rate (deaths/yr × 10⁻³) among males aged 50–69 versus income and air pollution level in the section of Buffalo in which they live. From Ref. 28. TSP means total suspended particulates.

Rotterdam, Oslo, and others, all indicating strong correlations with air pollution levels.

If we take these analyses at face value, we conclude that there are about 50,000 deaths in the United States each year due to air pollution,[3,30] and about 20,000 of these are due to coal burning. However, there is a tendency in the involved scientific community to consider this number to be too large. My estimate from extensive reading and fairly close contact with the scientists working on the problem is that an average opinion among them would be 10,000 deaths per year from coal burning. A little over half of our electricity is produced by coal burning, and a little over half of our coal is burned to produce electricity; therefore if all of our electricity were derived from coal, the resultant air pollution would cause 10,000 deaths per year.

Since there are about 2 million total deaths per year in the United States, this means that 0.5% of all Americans die prematurely due to coal-burning air pollution. It is typically estimated that the average victim loses about 7 years of life expectancy; hence the average LLE for all Americans due to this risk is $(0.005 \times 7 =) 0.035$ years, or 13 days.

As is evident from the preceding discussion, this number is quite uncertain. It could be ten times larger or ten times smaller. Antinuclear activists constantly use the uncertainty in radiation risks from nuclear power to imply that things may be much worse than usually estimated; they get a lot of mileage out of this ploy, but the uncertainties in health effects of air pollution from coal burning are very much larger.

In any case, there can be little question but that the LLE from air pollution due to coal burning is considerably larger than that due to all the radiation and accidents from the nuclear industry even if the claims of the antinuclear adherents are accepted—13 days versus 1.5 days. That is the basic reason for the conclusion of the many studies cited in Chapter 1 that nuclear power is safer than coal burning.

While the evidence for health effects from air pollution is undeniable, reaching an understanding of them has proved to be a very difficult task. Historically, the pollutants most easily and therefore most frequently measured were sulfur oxides and suspended particulate matter; thus nearly all correlation studies were based on them. Until ten years ago, it was widely assumed that these are the materials actually responsible for the health damage, but animals exposed to very high levels of these for long time periods showed no ill effects. Moreover, men occupationally exposed to 100 times the normal outdoor levels of sulfur oxides from refrigeration sources, oil refineries, and pulp and paper mills were not seriously affected. There are occupational settings where suspended particulate levels are 100 times the outdoor average, again without important health effects.[32] In response to these findings, in the mid-1970s there was a strong trend toward designating "sulfate particulates," resulting from chemical reactions of sulfur oxides with other chemicals in air, as the culprit, and some still continue to support that viewpoint; but animals show no ill effects even from prolonged exposure to relatively high levels of sulfate particulates.

One could concentrate on other components of air pollution as possible sources of the health effects, and there are plenty of candidates: nitrogen oxides, oxidants, polycyclic organic matter, and toxic metals like lead, mercury, cadmium, arsenic, or beryllium, some of which tend to attach to surfaces of the particulates. All of these have been studied[33]; all have been found to cause health effects but there is no evidence or strong body of opinion that any one of these is the major culprit in air pollution. One could also conjecture that health effects arise from complicated interactions of several pollutants acting together. This is probably the case, but it would be very hard either to prove or disprove.

The greatest difficulty in trying to tie down causes is that air pollution doesn't kill healthy people in one fell swoop. It rather continuously weakens the respiratory and cardiovascular systems over many decades until a time is reached at an advanced age when these collapse under one added insult. This explains why there are no mortality effects on the occupationally exposed, on animals, and on college students used as volunteer subjects in controlled tests.[32,33]

Since we do not know what components of air pollution cause the health effects, it is impossible to know what pollution control technologies will be effective in averting them. The often-heard statement "we can clean up coal burning" involves a large measure of wishful thinking.

RISKS IN OTHER ENERGY TECHNOLOGIES[1]

Other than coal burning, the principal alternatives to nuclear power are oil, gas, solar energy, and conservation, i.e., reducing our energy use. We consider only their health impacts here, although these are by no means their most important drawbacks and some of their other problems are discussed elsewhere in this book. Burning oil causes some air pollution, although not nearly as much as coal burning; it is also responsible for fires, giving a total LLE of 4 days.[1] Natural gas causes a little air pollution and some fires, but also kills by explosions and asphyxiation, giving an LLE of $2\frac{1}{2}$ days. The principal health impact of solar energy* is in the coal that must be burned to produce the vast quantities of steel, glass, and concrete required to emplace the solar collectors; this is about 3% of the coal that would be burned to produce the same energy by direct coal burning,[4] so the health effects are 3% of those of the latter, or an LLE of 0.4 days if we obtained all of our electricity from the sun. This makes solar electricity far more dangerous to health than nuclear energy according to estimates by most scientists. The quantities of material used in other technologies are many times less than those required for solar technologies.

All electrical energy technologies bring with them the risk of electrocution, which now has an LLE of 5 days for the average American. Note that this is far higher than the effects of generating nuclear electricity even if we accept the estimates of the antinuclear group. If solar electricity is generated and power conditioned in homes, it would probably multiply this effect manyfold.

The final alternative to nuclear power is conservation, doing without so much energy. Many people think this is the safest strategy, but it is probably

---

* This applies to short term effects. Long term effects[34] will be discussed in Chapter 5.

by far the most dangerous. One energy conservation strategy is to use smaller cars, but we have shown that this has an LLE of 50 days, many times that of any other energy technology. The danger would be somewhat reduced if everyone used small cars, but most fatal accidents are from collisions with fixed objects, like trees or walls, and collisions with large trucks and buses are also important; the risk they pose is greatly increased by using small cars.

If fuel conservation doubles the amount of bicycling, the resulting LLE is 10 days for the average American. Bicycles are far more dangerous than small cars, and motorcycles are in between, both in danger and in fuel conservation. Another energy conservation strategy is to seal buildings more tightly to reduce the escape of heat, but this traps unhealthy materials inside. One such material is radon, a naturally radioactive gas which is constantly percolating out of the ground and out of plaster and other building materials. Tightening buildings to reduce air leakage in accordance with government recommendations would give the average American an LLE of 24 days due to increased radon exposure,[35] making conservation by far the most dangerous energy strategy from the standpoint of *radiation* exposure!

Still another conservation strategy is to reduce lighting. Falls now give us an LLE of 39 days; thus if reduced lighting causes 5% more falling, it has an LLE of 2 days. If reduced lighting increases the number of murders by 5%, this would give an additional LLE of 4.5 days. Most motor vehicle accidents occur at night, and if reduced road lighting increases their number by even 2%, this gives an additional 4 days of LLE. Note that each of these LLE is larger than that due to nuclear power even if the estimates of the antinuclear activists are accepted.

An important *potential* danger in overzealous energy conservation is that it may reduce our wealth by suppressing economic growth. Our electricity supplies are adequate at present, but these are times of economic stagnation. If and when economic growth returns, as it must if we are to provide the one million-plus new jobs required each year for the rest of this century, more electricity will be needed to build and operate new industrial plants. If it is not available, these plants cannot materialize, and it will be too late to do much about it as power plants take many years to build. But power plant construction starts have been minimal for the past several years due to pressures for energy conservation.

We have already pointed out that the LLE from being poor in the United States is several years. The poor in a wealthy country, however, still get many health benefits from the national wealth—free medical care, food subsidies, relief payments, and so on. In countries where the whole nation is poor, the situation is much less favorable. In well-to-do countries like the

United States, Western Europe, Australia, and Japan, life expectancy is about 71 years, whereas life expectancy in a sample of other countries is: 68 years in Poland and Rumania, 61 years in Mexico, 55 years in Turkey, 45 years in India, and 32 years in some African countries.[36] Lest these differences be ascribed to racial factors, it should be noted that Japanese have ten years more life expectancy than other East Asians, and blacks in the United States have more than 20 years longer life expectancy than African blacks. The history of white versus black life expectancy in the United States is illuminating here: in 1900, there was a 20 year difference (52 years versus 32 years) whereas it is now reduced to 6 years, reflecting the improving socioeconomic status of blacks (but also showing that much progress remains to be made in that regard).

From these statistics it is evident that reduced national wealth can have disastrous impacts on health, amounting to hundreds of days of LLE. The greatest potential risk in overzealous energy conservation is that it will lead us down that thorny path toward becoming a poor nation.

Since we have estimated the LLE from coal burning as only 13 days, that technology is much less risky than doing without the electricity we need. Coal burning is an acceptable method for generating electricity, but is far inferior to nuclear energy for that purpose.

Some people seem to believe that reducing our energy usage is unavoidable because we are running out of fuel, but that is most definitely *not* the case. We will show later that nuclear fission can easily provide all the energy the world will ever need without any increase in fuel costs. No one favors *wasting* energy; waste is bad by definition. But there is no long-term reason to deny ourselves any convenience, comfort, or pleasure that energy can bring us, as long as we are willing to pay a fair market price for it.

SPENDING MONEY TO REDUCE RISK[37]

Another aspect of understanding risk is to consider what we are doing— or deciding not to do—to reduce our risks. Surely it is unreasonable to spend a lot of money to reduce one risk if we can much more cheaply reduce a greater risk but are not doing so.

It may seem immoral and inhumane even to consider the monetary cost of saving lives, but the fact is that a great many of our risks can be reduced by spending money. A few years ago, air bags were offered as optional safety equipment on several types of automobiles, but they are no longer offered because not enough people were willing to buy them. They were proven to be effective and safe—an estimated 15,000 lives per year would be saved if they were installed in all cars.[38] There is no discomfort or inconvenience connected with them. They have only one drawback—they cost money. Apparently Americans did not feel that it was worth the money to reduce their risk of being killed or injured in an automobile accident.

There is a long list of other automobile safety features we can buy—premium tires, improved lights, and rear window de-icers, to name a few. We can spend money on frequent medical examinations, we can use only the best and most experienced doctors, we can buy elaborate fire protection equipment for our homes, we can fly and rent a car at our destination rather than drive on long trips, we can move to safer neighborhoods—the list is endless. Each of these also costs money. In this section we consider how much it costs to save a life by spending money in various ways. In some cases, where personal effort and time are also required, a reasonable monetary compensation for these will be added to the cost.

As an example, getting a Pap smear to test for cervical cancer requires making an appointment and spending a few minutes at the doctor's office, but most women would be willing to do equivalent chores for a payment of $10. A Pap test costs* about $20, so we add the $10 for time and effort and take the total cost to be $30. Each annual Pap test has one chance in 3,000 of saving a woman's life[39]; thus for every 3000 tests, costing (3000 × $30 =) $90,000, a life is saved. The average cost per life saved is then $90,000. About 50% of U.S. women of susceptible age now have regular Pap tests.[40] If you are among the other 50%, you are effectively deciding that saving your life is not worth $90,000.

This example is taken from a study[37] completed a few years ago in which all costs are given in 1975 dollars. Some other calculational details will be given in the Appendix, but here we will quote some of the results of that study. Since the Consumer Price Index (CPI) increased by 75% from 1975 to 1982, the reader might apply that factor to the costs we give. Medical costs increased more rapidly than the CPI.

If there were a smoke alarm in every home,[41] it is estimated that 2,000 fewer people would die each year in fires. Even with a generous allowance for costs of installation and maintenance, this works out to a life saved for

---

* This example was originally given in 1975 dollars; costs have been doubled to account approximately for inflation. Note that a typical visit to a gynecologist involves much more than a Pap test. Waiting time is not included as it could be used profitably or recreationally (e.g., reading).

every $60,000 spent, but only about 25% of American homes have smoke alarms.

On the other hand, a great many Americans purchase premium tires to avert the danger of blowouts. If everyone did, this would cost an aggregate of about $5 billion per year and might avert nearly all of the 1,800 fatalities per year that result from blowouts, a cost of nearly $3 million per life saved. Many Americans buy larger cars than they need in order to achieve greater safety, which costs something like $6 million per life saved.[37]

There is clearly no logical pattern here. It is not that some people feel that their life is worth $6 million while others do not consider it to be worth even $90,000—there are undoubtedly many women who buy larger cars for safety reasons but skip their regular Pap test. And there are millions of Americans who purchased premium tires with their new cars but did not order air bags, even though the air bags are ten times more cost effective. The problem is that the American consumer does not calculate cost effectiveness. His actions are often governed by advertising campaigns, salesmanship, peer group pressures, and a host of other psychological and sociological factors.

But what about the government? We pay a large share of our income to national, state, and local governments to protect us. The government can hire scientists or solicit testimony from experts to determine risks and benefits, or even to develop new methods for protecting us; they have the financial resources and legal power to execute a wide variety of health and safety measures. Yet how consistently has the government functioned in this regard?

First let us consider cancer-screening programs. The government could implement measures to assure that much higher percentages of women get annual Pap smears; this has been done in a few cities like Louisville, Toledo, Ostfold (Norway), Aberdeen (Scotland), and Manchester (England). Ninety percent participation was achieved by such measures as sending personal letters of reminder or visits by public health nurses.[42] Such measures would involve added costs, but tests would be cheaper when done in a large-scale program[39]—a Mayo Clinic program did them for $3.50 in the 1960s and a British program did them for $2 apiece in 1970—so thousands of lives could be saved each year at a cost below $50,000 each.

There are several other cancer-screening programs that could be implemented. Fecal blood tests can detect cancer of the colon or rectum for as little as $10,000 per life saved.[43] Many more of these cancers could be detected in men aged 50–65, the most susceptible age, by annual proctoscopic examinations,[44] saving a life for every $30,000 spent, but only one in eight men of this age now get such examinations. Lung cancer can be detected by sputum cytology and by X-ray examination; the Mayo Clinic has been running such a program for heavy cigarette smokers which saves a life for every $65,000 spent,[45] and two programs in London reported success at less than half of that cost.[46] Nevertheless, only a small fraction of American adults are screened each year, and there has been little enthusiasm for large government-sponsored programs. Estimated costs per life saved are similar for breast cancer,[47] the leading cause of cancer death in women which is readily curable with early detection, but only half of all women are serviced and again there are no large government programs.

Testing for high blood pressure has almost become a fad in this country, but the problem goes beyond detection. Treatment is quite effective but since the condition is not immediately life threatening, many people ignore it. A well-organized treatment program would save a life for every $75,000 invested,[48] with half of that cost compensating patients for their inconvenience, but such programs have not been developed.

An especially effective approach to saving lives with medical care is with mobile intensive care units (MICU),[49] well-equipped ambulances carrying trained paramedics ready to respond rapidly to a call for help. About one-third of all deaths in the U.S. are from heart attacks (one-third of these are in people less than 65 years old) and two-thirds of these deaths occur before the patient reaches the hospital. The MICU was originally conceived as a method for combating this problem by providing rapid, on-the-scene coronary monitoring and defibrillation services, but they have now been expanded to provide treatment for burns, trauma, and other emergency conditions. Experience in large cities has shown that they save lives for an average cost of about $12,000; consequently every large city has them. However, for smaller towns the cost goes up. When it reaches $30,000 per life saved—the cost for a town with a population of 40,000—it is often considered too expensive.[50] In effect, it is decided that saving a life is not worth more than $30,000.

In summary of our medical examples, there are several available programs that could save large numbers of lives for costs below $50,000 each, and many more for costs up to $100,000 per life saved. These, of course, are American lives with some chance that they may be our own. Lifesaving overseas is much cheaper. Food to underdeveloped countries like India could effectively save one life for every $5,000 spent,[51] and there is an immunization program in Indonesia that can save 300,000 lives at a cost of $30 million, or $100 per life saved.[52]

But health care is not our government's only means of spending money to save lives. Over 35,000 Americans die in automobiles each year as a result of collisions, and over a million are seriously injured even though there are many ways in which this toll could be reduced by investing money in highway or automobile safety devices.

To some extent this has been done. A number of new safety devices in automobiles, like collapsible steering columns and soft dashboards, were mandated by law between 1965 and 1974; a study by the U.S. General Accounting Office[53] indicates that they have saved a life for every $140,000 spent. However, this is apparently about as high as we are willing to go, and the program has ground to a halt. In 1970–1973, 16 new safety measures in automobiles were mandated, but there have been hardly any since that time. As noted previously, an air bag requirement which would cost $300,000 per life saved has not been implemented.

There are lots of highway construction measures that could save lives. For example, about 6,000 Americans die each year in collisions with guardrails, and there are guardrail construction features that could absorb much of the impact and thereby save most of those lives.[53] But let us now get down to costs per life saved.

The "National Highway Safety Needs Report,"[55] published in 1976, represented a federal government effort to estimate the cost effectiveness of various highway safety measures. It found that guardrail improvements in selected locations could save over 300 lives per year at a cost of $34,000 each. Improvements in regulatory and warning signs would save even more lives at a similar unit cost. Other measures that would each save hundreds of lives per year at costs of under $50,000 each include using construction techniques which improve skid resistance, and designing safer bridge rails and parapets. Using standard techniques for avoiding wrong-way entrance onto freeways would save about 80 lives per year at that cost.

If we were willing to get into the $100,000 per life saved cost range, we could save 680 lives per year with impact-absorbing devices at critical roadside points, and 325 lives per year by use of breakaway sign and lighting supports rather than rigid supports for these on high-speed highways.

It is interesting to note that the cost per life saved becomes too high much sooner for highway safety devices than for automobile safety; perhaps this is because in highway safety the government spends its money, whereas in automobile safety it only requires citizens to spend their own money.

It is estimated[56] that high school courses in driver education avert about 6,000 fatalities per year and cost less than $100,000 per life saved, even if we include a $50 payment to each student for his time and trouble. Yet there are recent indications that programs are being cut back to save on costs.

Before leaving the area of traffic safety it should be pointed out that there are 40 serious injuries for every fatality in traffic accidents.[2] The measures we have discussed would reduce the former as well as the latter. We have therefore erred on the high side in charging all of the costs to lifesaving; the costs per life saved are lower than given in the above discussion.

We have seen that some of our governmental agencies are passing up opportunities for saving lives at costs below $50,000, and they are rarely willing to spend over $200,000. But this does not apply to the Environmental Protection Agency in protecting people from pollution. In its regulation dealing with air pollution control equipment for coal-burning power plants, it requires installation of sulfur scrubbers in nearly all cases, which corresponds to spending an average of $500,000 per life saved. Moreover, air pollution usually affects old people with perhaps an average of seven years of remaining life expectancy, whereas automobile accidents generally kill young people whose life expectancy averages 40 years.

Where radiation is involved, EPA hastens to go much further. Radium is a naturally occurring element that is found in all natural waters. This has always been so, and it always will be so. However, EPA is now requiring that in cases where radium content is abnormally high in drinking water, special measures be taken to remove some of it. This, it estimates, corresponds to spending $2.5 million per life saved.[57]

Nonetheless EPA is not alone in being willing to spend heavily from the public purse to reduce radiation exposure. The Office of Management and Budget recommended in 1972 that nuclear reactor safety systems be installed where they can save a life for every $8 million spent.[58] Not to be outdone, the Nuclear Regulatory Commission also requires an $8 million expenditure per life saved in controlling normal emission of radioactivity from nuclear power plants.[59] But NRC furthermore, has special rules for special substances; regulations on emissions of radioactive iodine correspond to spending $100 million per life saved.

In the matter of spending money to save lives from radiation exposure, however, a new champion has recently come on the scene—nuclear waste. This will be discussed in the next chapter, where it will be shown that we are spending hundreds of millions and even billions of dollars per life saved in this area.

A summary of the costs per life saved developed here, plus others from Ref. 36, is shown in Table 2. Also shown there are the costs per 20 years of life expectancy added. Clearly, there are enormous wastes of money and of lives from the illogical inconsistency in our spending for life saving. Thousands of Americans are dying needlessly each year for lack of money spent on medical programs and traffic safety, while the money that could save them is being spent to save a single life from other hazards.

TABLE 2

COST PER FATALITY AVERTED (1975 DOLLARS) IMPLIED BY VARIOUS SOCIETAL
ACTIVITIES (LEFT COLUMN) AND COST PER 20 YEARS OF ADDED LIFE EXPECTANCY
(RIGHT COLUMN)

Item	$ per fatality averted	$ per 20-yr life expectancy
Medical screening and care		
Cervical cancer	25,000	13,000
Breast cancer	80,000	60,000
Lung cancer	70,000	70,000
Colorectal cancer		
Fecal blood tests	10,000	10,000
Proctoscopy	30,000	30,000
Multiple screening	26,000	20,000
Hypertension control	75,000	75,000
Kidney dialysis	200,000	440,000
Mobile intensive care units	30,000	75,000
Traffic safety		
Auto safety equipment—1966–1970	130,000	65,000
Steering column improvement	100,000	50,000
Air bags, driver only	320,000	160,000
Tire inspection	400,000	200,000
Rescue helicopters	65,000	33,000
Passive three-point harness	250,000	125,000
Passive torso belt-knee bar	110,000	55,000
Driver education	90,000	45,000
Highway construction maintenance practice	20,000	10,000
Regulatory and warning signs	34,000	17,000
Guardrail improvements	34,000	17,000
Skid resistance	42,000	21,000
Bridge rails and parapets	46,000	23,000
Wrong-way entry avoidance	50,000	25,000
Impact absorbing roadside devices	108,000	54,000
Breakaway sign, lighting posts	116,000	58,000
Median barrier improvement	228,000	114,000
Clear roadside recovery area	284,000	142,000
Miscellaneous nonradiation		
Expanded immunization in Indonesia	100	50
Food for overseas relief	5,300	2,500
Sulfur scrubbers in power plants	500,000	1,000,000
Smoke alarms in homes	60,000	40,000
Higher pay for risky jobs	260,000	150,000

Continued

TABLE 2 *(Continued)*

Item	$ per fatality averted	$ per 20-yr life expectancy
Coalmine safety	22,000,000	13,000,000
Other mine safety	34,000,000	20,000,000
Coke fume standards	4,500,000	2,500,000
Air Force pilot safety	2,000,000	1,000,000
Civilian aircraft (France)	1,200,000	600,000
Radiation related activities		
Radium in drinking water	2,500,000	2,500,000
Medical X-ray equipment	3,600	3,600
OMB[a] guidelines	7,000,000	7,000,000
Radwaste practice—general	10,000,000	10,000,000
Radwaste practice—[131]I	100,000,000	100,000,000
Defense high-level waste	200,000,000	200,000,000

[a] OMB is the Office of Management and Budget which monitors U.S. Government spending.

We have already discussed in Chapter 1 the reason for this deplorable situation. Government officials must be responsive to public concern, so if the public is much more concerned about dangers of radiation than about highway or medical hazards, the government will spend much more to protect them from the former than from the latter. In a democracy like ours, the public calls the shots.

The problem is in the public's perception of where it needs protection, in its failure to understand and quantify risk and to put various risks in proper perspective. I blame this misunderstanding on the media. If they paid more attention to the real risks in our society rather than concentrating on the imagined risks of nuclear power, thousands of lives and billions of dollars could be saved every year in the United States.

APPENDIX

For those readers who are interested, we demonstrate here how to calculate some of the results quoted in this chapter.

First, we calculate the LLE from reactor accidents according to the Nuclear Regulatory Commission Study[20] which estimates one meltdown per

20,000 reactor-years of operation, and an average of 400 fatalities per meltdown. All U.S. electricity derived from nuclear power plants would require about 250 such plants, giving us 250 reactor-years of operation each year. We would therefore expect a meltdown every (20,000/250 =) 80 years on an average. The average fatality rate is then (400/80 =) 5 per year. If the U.S. were to maintain its present population for a long time there would be about 3 million deaths each year, so (5/3 million =) 1.7 out of every million deaths would be due to nuclear accidents. Victims of nuclear accidents lose an average of 20 years of life expectancy (cancers from radiation usually develop 15 to 50 years after exposure), giving the average American an LLE = ($1.7 \times 10^{-6} \times 20$ =) 34×10^{-6} years; multiplying this by 365 days/year $\times$ 24 hours/day $\times$ 60 minutes/hour gives LLE = 18 minutes as quoted previously.

The UCS estimates[25] are one meltdown every 2000 reactor-years, with an average of 5000 deaths per meltdown. Since these numbers are, respectively, 10 and 12.5 times higher than the NRC estimates,[20] the LLE is larger by a factor of $10 \times 12.5 = 125$. Multiplying this by 18 minutes gives 2250 minutes or 1.5 days. Alternatively, one could go through the entire calculation in the previous paragraph.

We next calculate the LLE from being exposed to 0.2 mrem/yr, which is a lifetime exposure of (71 years $\times$ 0.2 mrem/yr =) 14 mrem. In Chapter 2 we found the cancer risk to be one chance in 8 million per millirem. Multiplying this by 14 mrem gives a lifetime cancer risk of 1.7 chances in a million, the same as the NRC estimate from reactor accidents. The rest of the calculation is the same as for the latter, and therefore gives the same result, an LLE of 18 minutes.

This corresponds to an LLE of 1.2 minutes per millirem (18 min/14 mrem). The average exposure in the Three Mile Island accident to people living in that area was 1.2 mrem,[60] so their LLE was (1.2 $\times$ 1.2 $\simeq$) 1.5 minutes.

The comparisons of risks are based on the ratio of LLE. For example, if one pound of added weight gives an LLE of 30 days while the UCS estimate gives an LLE from reactor accidents of 1.5 days, gaining one pound is $30 \div 1.5 = 20$ times more dangerous, or gaining 1/20 of a pound must be equally dangerous. Multiplying by 16 ounces per pound gives 0.8 ounces as the weight gain giving equivalent risk.

As an example of calculating the cost per life saved, consider the use of air bags. According to Allstate Insurance Company,[38] an air bag reduces the drivers' mortality rate by 1.4 deaths per hundred million miles driven. Therefore, if a car is driven 50,000 miles, its probability of saving a life is

1.4 × (50,000/100,000,000) = 1/1500, or one chance in 1500. This air bag would cost about $200, so the cost per life saved is $200 divided by 1/1500, or $300,000. Another way of saying this is that for every 1500 cars equipped with an air bag, an average of one life would be saved; the cost would then be 1500 × $200 = $300,000 to save one life.

As another example, consider the Nuclear Regulatory Commission regulation requiring installation of any equipment in a nuclear plant that will reduce the total exposure to all members of the public* by 1 mrem per dollar spent.[59] It was shown in Chapter 2 that we can expect one fatal cancer for every 8 million millirem; thus the regulation requires spending $8 million for each death averted.

REFERENCE NOTES

1. B. L. Cohen and I. S. Lee, "A Catalog of Risks," *Health Physics,* **36,** 707 (1979). This paper cites a large number of references to original sources of information.
2. B. L. Cohen, "Perspective on Occupational Risks," *Health Physics,* **40,** 703 (1981).
3. United Nations Scientific Committee on the Effects of Atomic Radiation, "Sources and Effects of Ionizing Radiation," United Nations, New York (1977).
4. B. L. Cohen, "Society's Valuation of Life Saving . . . ," *Health Physics,* **38,** 33 (1980), This paper cites a large number of references to original sources of information.
5. A. R. Karr, "Saga of the Air Bag," Insurance Institute for Highway Safety, Status Report, November 30 (1976).
6. U. S. Bureau of Census, Vital Statistics of the United States (published annually).
7. NCHS (National Center for Health Statistics), "Mortality from Selected Causes by Marital Status," NCHS Series 20, No. 8a,b, December (1970).
8. H. A. Kahn, "The Dorn Study of Smoking and Mortality Among U.S. Veterans," in National Cancer Institute Monograph 19, W. Haenszel (ed.), U.S. Department of Health, Education and Welfare (1966); E. C. Hammond, "Smoking in Relation to the Death Rates of One Million Men and Women," in National Cancer Institute Monograph 19 W. Haenszel (ed.), U.S. Department of Health, Education and Welfare (1966); PHS (U.S. Public Health Service), "The Health Consequences of Smoking—A Public Health Service Review—1967" (1967).
9. Metropolitan Life Insurance Company, Statistical Bulletin, Feb., 1960; Mar., 1960; Society of Actuaries, Build and Blood Pressure Study (1959).
10. B. L. Cohen, "Body Weight as an Application of Energy Conservation," *American Journal of Physics,* **45,** 867 (1977).

* For example, if 100 people are exposed to 1 mrem each, the total exposure is 100 mrem.

11. U.S. Bureau of Census, "Statistical Abstract of the United States—1975" (1975).
12. PHS (U.S. Public Health Service), Vital Statistics—Special Reports, Vol. 53, No. 4 (1962); Metropolitan Life Insurance Company Statistical Bulletin, Socioeconomic Mortality Differentials, January (1975).
13. Registrar General, *1961 Occupational Mortality Tables* (Her Majesty's Stationery Office, London, 1971).
14. E. M. Kitagawa and P. M. Hauser, " Education Differentials in Mortality by Cause of Death in the U.S.," *Demography*, **5**, 335 (1968).
15. Insurance Institute of Highway Safety, Vol. 10, No. 12 (9 July, 1975); U.S. Department of Transportation, "The Car Book," Document DOT HS 805 580 (1981).
16. Metropolitan Life Insurance Co., Statistical Bulletin, "Mortality from Accidents by Age and Sex," May (1971).
17. NCHS (National Center for Health Statistics), "State Life Tables, 1969–71," DHEW Publication No. (HRA) 75-1151 (1975).
18. R. Wilson and J. Chase, "Problems Associated with Intercomparisons of Risk," in First International Conference on Health Effects of Energy Production, Atomic Energy of Canada Ltd. Document AECL 6958 (1979).
19. Metropolitan Life Insurance Company, Statistical Bulletin, "Hazardous Occupations and Avocations," March (1974); Metropolitan Life Insurance Company, Statistical Bulletin, "Fatalities in Motor Vehicle Racing," December (1976).
20. Registrar General, *Occupational Mortality* (Her Majesty's Stationery Office, London, 1978); Society of Actuaries, "Occupational Studies," Chicago (1967); U.S. Public Health Service, "Vital Statistics," Special Reports 53 (4), (1962).
21. NRC (U.S. Nuclear Regulatory Commission), "Reactor Safety Study," WASH-1400, NUREG-75/014 (1975).
22. NSF (National Science Foundation) Science and Technology Policy Office Report, "Chemicals and Health" (1973).
23. FDA (U.S. Food and Drug Administration), in *Federal Register, 42* (73), 20001 (15 April 1977); also press release.
24. Calculations by author unpublished.
25. Union of Concerned Scientists (UCS), "The Risks of Nuclear Power Reactors," Cambridge, Massachusetts (1977).
26. United Nations Scientific Committee on the Effects of Atomic Radiation, "Sources and Effects of Ionizing Radiation," United Nations, New York (1977).
27. C. Starr, "Social Benefits vs. Technological Risks," *Science, 165,* 1232 (1969).
28. R. Wilson and W. I. Jones, *Energy, Ecology, and the Environment* (Academic Press, 1979); M. Glasser and L. Greenburg, "Air Pollution and Mortality and Weather, New York City 1960–64," *Archives of Environmental Health, 22,* 334 (1971).
29. W. Winkelstein *et al.*, "The Relationship of Air Pollution and Economic Status to Total Mortality and Selected Respiratory System Mortality for Men," *Archives of Environmental Health, 14,* 162 (1967).

30. M. Chappie and L. Lave, "The Health Effects of Air Pollution: A Reanalysis," *Journal of Urban Economics*, **12**, 346, July (1981); T. D. Crocker, W. Schulze, S. Ben-David, and A. Kneese, "Methods of Development for Assessing Air Pollution Control Benefits," Volume I, "Economics of Air Pollution Epidemiology," U.S. EPA Report EPA-600/5-79-001a, U.S. EPA, Washington, D.C., (1979); S. Gerking, and W. Schulze, "What Do We Know About the Benefits of Reduced Mortality from Air Pollution Control?" *American Economic Review*, **71**(2), 228–234 (1981); J. J. Gregor, *Mortality and Air Quality, The 1968–1972 Allegheny County Experience* (Center for the Study of Environmental Policy, Pennsylvania State University, University Park, Pennsylvania 1976); R. J. Hickey, D. E. Boyce, R. C. Clelland, E. J. Bowers, and P. B. Slater, "Demographic and Chemical Variables Related to Chronic Disease Mortality in Man," Technical Report No. 15, Department of Statistics, University of Pennsylvania (1977); R. Koshal and M. Koshal, "Air Pollution and the Respiratory Disease Mortality in the United States—A Quantitative Study," *Social Indicators Research*, **1**: 263–278 (1974); L. B. Lave and E. P. Seskin, "Air Pollution and Human Health," *Science*, **169**, 723–733 (1970); L. B. Lave and E. P. Seskin, "An Analysis of the Association Between U.S. Mortality and Air Pollution," *Journal of the American Statistical Association*, **68**(6), 284–290 (1973); L. B. Lave and E. P. Seskin, *Air Pollution and Human Health* (Johns Hopkins University Press, Baltimore, 1977); L. B. Lave and E. P. Seskin, "Epidemiology, Causality and Public Policy," *American Scientist*, **67**,178–186 (1979); F. W. Lipfert, "Statistical Studies of Geographical Factors in Cancer Mortality: Association with Community Particulate Air Pollution," presented at the 71st Annual Meeting of the Air Pollution Control Association, Houston, Texas, June 25–30 (1978); W. Lipfert, "The Association of Human Mortality with Air Pollution: Statistical Analyses by Region, by Age, and by Cause of Death," Long Island Lighting Company (1978); B. C. Liu and E. S. H. Yu, "Mortality and Air Pollution Revisited," *Journal of the Air Pollution Control Association*, **26**(10), 968–971 (1977); R. Mendelsohn and G. Orcutt, "An Empirical Analysis of Air Pollution Dose–Response Curves", *Journal of Environmental Economics and Management*, **6**, 85–106 (1979); R. C. Schwing and G. C. McDonald, "Measures of Association of Some Air Pollutants, Natural Ionizing Radiation, and Cigarette Smoking with Mortality Rates," *The Science of the Total Environment*, **5**,139–169 (1976); S. Selvin, D. Merrill, L. Kwok, and S. Sacks, "Ecologic Regression Analysis and the Study of the Influence of Air Quality on Mortality," Lawrence Berkeley Laboratory preprint LBL-12217, Berkeley, California (1981); J. J. Seneca, P. Asch, and K. Brennan, "Mortality and Air Pollution in New Jersey," In 12th Annual Report—"Economic Policy Council and Office of Economic Policy," State of New Jersey, Dept. of the Treasury, New Jersey (1979); V. K. Smith, "The Measurement of Mortality—Air Pollution Relationships," *Environment and Planning*, **8**, 149–162 (1976); V. K. Smith, *The Economic Consequences of Air Pollution* (Ballinger Publishing Co., Cambridge, Massachusetts, 1976); T. J. Thomas, "An Investigation into

Excess Mortality and Morbidity Due to Air Pollution," Ph.D. dissertation, Purdue University, W. Lafayette, Indiana (1973).

31. R. W. Buechley, W. B. Riggan, V. Hasselblad, and J. B. VanBruggen, "SO₂ Levels and Perturbations in Mortality: A Study in the New York–New Jersey Metropolis," *Archives of Environmental Health*, **27**, 134–137 (1973); A. T. Gore and C. W. Shaddick, "Atmospheric pollution and mortality in the county of London," *British Journal of Social Medicine*, **12**, 104 (1958); H. Heimann, "Episodic Air Pollution in Metropolitan Boston: A Trial Epidemiological Study" *Archives of Environmental Health*, **20**,230 (1970); L. B. Lave and E. Seskin, *Air Pollution and Human Health*, (Johns Hopkins University Press, Baltimore, 1977); A. E. Martin and W. H. Bradley, "Mortality, Fog and Atmospheric Pollution: An Investigation During Winter of 1958–59," *Monthly Bulletin of Mining Health (London)*, **19**, 56–72 (1960); H. Schimmel and L. Greenberg, "A Study of the Relation of Air Pollution to Mortality, New York City 1963–1964," *Journal of the Air Pollution Control Association*, **22**, 606 (1973); H. Schimmel and T. J. Murawski, "The Relation of Air Pollution to Mortality," *Journal of Occupational Medicine*, **18**, 316 1976; J. A. Scott, "The London Fog of December, 1952," *The Medical Officer. (London)*, **109**, 250–252 (1953); R. E. Wyzga, "The Effect of Air Pollution on Mortality—A Consideration of Distributed Lag Models," *Journal of the American Statistical Association*, **73**, 463–472 (1978); A. E. Martin, "Mortality and Morbidity Statistics and Air Pollution," *Proceedings of the Royal Society of Medicine*, **57**, 969 (1964); T. A. Hodgson, "Short-Term Effects of Air Pollution on Mortality in New York City," *Environmental Science and Technology*, **4**, 589 (1970); M. D. Lebowitz, "A Comparative Analysis of the Stimulus–Response Relationship Between Mortality and Air Pollution–Weather," *Environmental Research*, **6**, 106 (1973); M. Glasser, and L. Greenburg, "Air Pollution and Mortality and Weather: New York City, 1960–64," *Archives of Environmental Health*, **22**, 334 (1971); S. E. Burgess and C. W. Shadick, "Bronchitis and Air Pollution," *Royal Society Health Journal*, **79**, 10 (1959); S. Mazumder and N. Sussman, "Relationships of Air Pollution to Health: Results from the Pittsburgh Study," Proc. of 74th Annual APCA Meeting, Philadelphia, June (1980); S. Mazumder, H. Schimmel, and I. Higgens, "Daily Mortality, Smoke, and SO₂ in London, England, 1959–72," Proc. of APCA, Pittsburgh Meeting, September (1980), p. 219.

32. U.S. Environmental Protection Agency, Air Quality Criteria for Particulate Matter and Sulfur Oxides (1981).

33. U.S. Environmental Protection Agency (EPA), "Air Quality Criteria for Oxides of Nitrogen" (1980); U.S. EPA, "Health Assessment Document for Polycyclic Organic Matter," EPA 600/9-79-008 (1979); U.S. EPA, "Health Assessment Document for Arsenic" (1980); U.S. EPA, "Air Quality Criteria for Lead" (1977).

34. B. L. Cohen, "Long Term Consequences of the Linear–No Threshold Dose–Response Relationship for Chemical Carcinogens," *Risk Analysis*, **1**, 267 (1981); P. D.

Moskowitz *et al.*, "Photovoltaic Energy Technologies: Health and Environmental Effects Document," Brookhaven National Lab. Report BNL-51284 (1980).

35. B. L. Cohen, "Health Effects of Radon from Insulation of Buildings," *Health Physics*, **39**, 937 (1980).

36. Population Reference Bureau, "1976 World Population Data Sheet," 1754 N St., N.W., Washington, D.C. (1978).

37. B. L. Cohen, "Society's Valuation of Life Saving in Radiation Protection and Other Contexts," *Health Physics*, **28**, 33 (1980).

38. A. R. Karr, "Saga of the Air Bag," *The Wall Street Journal*, Reprinted in *Insurance Institute for Highway Safety, Status Report*, **11**, 18 (November 30, 1976); L. M. Patrick, "Passive and Active Restraint Systems—Performance and Benefit–Cost Comparison," Society of Automotive Engineers Transactions, Paper 750389 (1975); W. Stork, "The Cost Effectiveness of International Vehicle Regulations," *Automotive Engineering*, March, 32 (1973).

39. L. Dickinson, "Evaluation of the Effectiveness of Cytologic Screening for Cervical Cancer," *Mayo Clinic Proceedings*, **47**, 550 (August 1972); J. B. Thorn, J. E. Macgregor, E. M. Russell, and K. Swanson, "Costs of Detecting and Treating Cancer of the Uterine Cervix in North-East Scotland in 1971," *Lancet*, 674 (March 22, 1975; R. J. Walton, "Cervical Cancer Screening Programs: The Walton Report," *Canadian Medical Association Journal*, **114**,(11) (June 5, 1976); R. N. Grosse, "Cost–Benefit Analysis of Health Service," *Annals of the American Academy*, **399**, 89 (1972); JEC (Joint Economic Committee), U.S. Congress, "The Analysis and Evaluation of Public Expenditure: The PPB System" (1969); S. E. Rhoads, "How Much Should We Spend to Save Lives?" *The Public Interest*, **Spring**, 74 (1978).

40. Gallup Organization (Princeton, New Jersey), January 1977. Survey sponsored by American Cancer Society (1977).

41. J. P. Ruchinskas, Private communications giving generally accepted value, General Electric Co., Housewares Division, Bridgeport, Connecticut (1978).

42. R. J. Walton, cf. Ref. 39.

43. M. Kristein, "Economic Issues in Prevention," *Preventive Medicine*, **6**, 252 (1977).

44. R. J. Bolt, "Sigmoidoscopy in Detection and Diagnosis of the Asymptomatic Individual," *Cancer*, **28**, 121 (1971); V. A. Gilbertson, "Proctosigmoidoscopy and Polypectomy in Reducing the Incidence of Rectal Cancer," *Cancer*, **34**,(3), Supplement, 936 (1974); R. N. Grosse, "Cost–Benefit Analysis of Health Service," *Annals of the American Academy*, **399**, 89 (1972).

45. R. S. Fontana, D. R. Sanderson, L. B. Woolner, W. E. Miller, P. E. Bernatz, W. S. Payne, and W. F. Taylor, "The Mayo Lung Project," *Chest* **67**, 511 (1975).

46. F. A. Nash, J. M. Morgan, and J. G. Tomkins, "South London Cancer Study," *British Medical Journal*, **2**, 715 (1968); G. Z. Brett, "Earlier Diagnosis and Survival in Lung Cancer," *Ibid.*, **4**, 260 (1969); K. R. Boucot, and W. Weiss, "Screening for Lung Cancer," *Journal of the American Medical Association*, **227**, 566 (1973); J. R. T. Colley, "Diseases of the Lung," *Lancet*, **ii**, 1125 (1974).

47. S. Shapiro, P. Strax, L. Venet, and W. Venet, "Changes in 5-year Breast Cancer Mortality in a Breast Cancer Screening Program," Seventh National Cancer Conference, Los Angeles, California (1972); M. Kristein, "Economic Issues in Prevention," *Preventive Medicine,* **6,** 252 (1977); L. M. Irwig, "Breast Cancer," *Lancet,* **ii,** 1307 (1974); R. N. Grosse, "Cost–Benefit Analysis of Health Service," *Annals of the American Academy,* **399,** 89 (1972).

48. J. Stamler, R. Stamler, W. F. Riedlinger, G. Algera, and R. H. Roberts, "Hypertension Screening of One Million Americans," *Journal of the American Medical Association,* **235,** 2299 (1976); J. Stokes and D. C. Carmichael, "A Cost–Benefit Analysis of Modern Hypertension Control," National Heart and Lung Inst., NIH, Bethesda, Maryland (1975).

49. R. Zeckhauser and D. Shepard, "Where Now for Saving Lives," *Law and Contemporary Problems,* **40**(5), 5 (1976); J. P. Acton, "Evaluating Public Programs to Save Lives: The Case of Heart Attacks," Rand Corp. Report R-950-RC, Los Angeles, California (1973).

50. J. Riordan, Private communication, Health Services Administration, Washington, D.C. (1979).

51. B. Ward, Paper presented at UNICEF New World Food Conference, Rome (1974); E. Egan, Private communication, Catholic Relief Services, New York (1977); see also Ref. 23.

52. H. N. Barnum, "An Economic Analysis of an Expanded Program of Immunization in Indonesia," U. of Michigan Report (1978).

53. General Accounting Office, "Effectiveness, Benefits and Costs of Federal Safety Standards for Protection of Passenger Car Occupants," Washington, D.C. (1976).

54. J. W. Sparks, "Development of an Effective Highway Safety Program," *Traffic Engineering,* January, 30 (1977).

55. U.S. Department of Transportation, "The National Highway Safety Needs Report" (1976).

56. R. C. Schwing, "Expenditures to Reduce Mortality Risk and Increase Longevity," General Motors Research Lab. Report GMR-2353-A (1978).

57. EPA (Environmental Protection Agency), Report EPA-570/9-76-00, Washington, D.C. (1976).

58. Office of Management and Budget (OMB), Circular A-94 (1972).

59. Code of Federal Regulations, Title 10, Part 50, Appendix I; U.S. Nuclear Regulatory Commission, (NRC) Regulatory Guides 1.109-1.113 (1976).

Chapter 5 / HAZARDS OF HIGH-LEVEL RADIOACTIVE WASTE—THE GREAT MYTH

The fourth major reason for public misunderstanding of nuclear power is a grossly unjustified fear of the hazards from radioactive waste. Even people whom I know to be intelligent and knowledgeable about energy issues have told me that their principal reservation about use of nuclear power is the disposal of radioactive waste. Often called an "unsolved problem," many consider it to be the Achilles' heel of nuclear power. Seven states now have laws prohibiting construction of nuclear power plants until the waste disposal issue is settled. On the other hand there is general agreement among those scientists involved with waste management that radioactive waste disposal is a rather trivial technical problem. Having it as one of my principal research specialties over the past seven years, I am firmly convinced that this radioactive waste from nuclear power operations represents less of a health hazard than the waste from any other large technological industry. Clearly there is a long and complex story to tell.

A FIRST PERSPECTIVE

What is this material that is so controversial? As we know from elementary physical science courses, matter can be neither created nor destroyed. When fuel is burned to liberate energy, the fuel doesn't simply disappear—it is converted into other forms which we refer to as "waste." This is true whether we burn uranium or coal or anything else. For nuclear fuels, this residue, called "high-level waste," has been the principal source of concern to the public.

As an initial perspective, it is interesting to compare nuclear waste with the analogous waste from a single large coal-burning power plant. The largest component of the coal burning waste is carbon dioxide gas, produced at a rate of 500 lb every second, 15 tons every minute. It is not a particularly dangerous gas but there is a great deal of concern that the tremendous quantities of it being generated in burning coal, oil, and gas will have important effects on world climate. It could raise the sea level to flood out coastal cities like New York, Miami, New Orleans, and Los Angeles, and convert our Midwestern grain belt into a dust bowl.[1] The most abundant *dangerous* gas emitted in coal burning is sulfur dioxide, discharged at a rate of a ton every five minutes. According to a National Academy of Sciences study commissioned by a U.S. Senate Committee,[2] the annual releases from a single plant results in 25 deaths, 60,000 cases of respiratory disease, and $25 million in property damage. Another type of gaseous pollutant from coal burning is nitrogen oxides, best known as the principal pollutant from automobiles—the reason for cars having expensive pollution control equipment and requiring lead-free gasoline. A single plant emits as much nitrogen oxide as 200,000 automobiles.[3]

Then there is the smoke, which consists of tiny solid particles. There is a widespread impression that the smoke from coal burning has been largely eliminated, but this is true only of the large particles that provide visible dirt. The situation is much less favorable regarding the smaller particles that are far more harmful because they can get past the body's defenses and reach the deep lung.[4] Another class of pollutant released in the burning of coal is "polycyclic hydrocarbons," a type of chemical that can cause cancer and genetic defects in later generations; the best known of these is benzpyrene, which is believed to be the principal cancer-causing agent in cigarette smoke. Then there is the ash, the bulk solid material produced at a rate of 1000 pounds per minute,[5] which is, in its disposal, responsible for some very difficult environmental problems, and for some serious long-term health effects to be discussed later. And finally there are uranium and thorium, naturally radioactive materials which serve as sources of radon gas with health effects, to be discussed in the next chapter, exceeding those of all radioactivity released from nuclear plants.[6]

The waste from a nuclear plant is different from these coal-burning wastes in two very spectacular ways. The first is in the quantities involved: the nuclear waste is five million times smaller by weight and billions of times smaller by volume. The nuclear waste from one year of operation weighs about 1½ tons[7] and would occupy a volume of half a cubic yard, which means that it would fit under an ordinary card table with lots of room to spare. Since the quantity is so small, it can be handled with a care and sophistication that is completely out of the question for the millions of tons of waste spewed out annually from our analogous coal-burning plant.

The second spectacular difference is that the nuclear wastes are *radioactive,* providing a health threat due to the radiation they emit, whereas the principal danger to health from coal wastes arise from their chemical activity. This does not mean that the nuclear wastes are more hazardous—on most comparison bases the opposite is true. For example, if all the air pollution emitted from a coal plant in one day were inhaled by people, 1½ million people could die from it,[8] which is ten times the number that could be killed by ingesting or inhaling the waste produced in one day by a nuclear plant.[9]

This is obviously an unrealistic comparison since there is no way in which all of either waste could get into people. A more realistic comparison would be on the basis of simple, cheap, and easy disposal techniques. For coal burning this would be to use no air pollution control measures and simply release the wastes without inhibition. This is not much worse than is done today since the smoke abatement techniques, which are the principal pollution control on most coal-burning plants, contribute little to health protection. The consequences of release of air pollution, as given above, is 25 fatalities per year from each plant. Some other comparably serious consequences will be considered later in this chapter.

For nuclear waste, a simple, quick, and easy disposal method would be to convert the waste into a glass*—a technology that is well in hand—and simply drop it into the ocean at random locations.[10] No one can claim that we don't know how to do that! With this disposal, the waste produced by one power plant in one year would eventually cause an average total of 0.6 fatalities, spread out over many millions of years, by contaminating seafood.

---

*Glass is a mixture of chemicals, and any number of chemicals in the proper form can be included in the mixture. The radioactive materials are converted into this form, thus becoming part of the glass.

Incidentally, this disposal technique would do no harm to ocean ecology. In fact if all the world's electricity were produced by nuclear power and all the waste generated for the next hundred years were dumped in the ocean, the radiation dose to sea animals would never be increased by as much as one percent above its present level from natural radioactivity.

We thus see that if we compare the nuclear and coal wastes on the basis of cheap, simple, and easy disposal techniques, the coal wastes are $(25 \div 0.6 =)$ 40 times more harmful to human health than nuclear waste. This treatment ignores some of the long-term health effects of coal burning to be discussed later, which increase this ratio to well over a hundred.

Probably the fairest comparison between the dangers of nuclear and coal wastes is to consider the actual technology to be used for each. We will do that later in this chapter, and it will be shown that on that basis, the nuclear wastes are thousands of times less harmful.

Another, and very different, way of comparing the dangers of nuclear and coal waste is on the basis of how much they are changing our exposures to toxic agents. The typical level of sulfur dioxide in the air of American cities is ten times higher than natural levels, and the same is true for the principal nitrogen oxides.[11] For cancer-causing chemicals the ratio is much higher.[12] These are matters that might be of considerable concern in view of the fact that we really do not understand the health effects of these agents very well. For radioactive waste from a flourishing nuclear industry, on the other hand, radiation exposures would be increased by only a fraction of one percent above natural levels.[13]

Another basis for comparison between the wastes from nuclear and coal burning is the "margin of safety" or, how much average exposures are below those for which there is direct evidence for harm to human health. For coal wastes, this is typically a factor of ten. There have been several air pollution episodes in which there were hundreds of excess deaths with sulfur dioxide levels about 1000 micrograms per cubic meter ($\mu g/m^3$),[4] whereas levels around 100 $\mu g/m^3$, just ten times less, are quite common in American cities.[14] For radiation, on the other hand, exposures expected from hypothetical problems with nuclear waste are in the range of 1 mrem or less, thousands of times below those for which there is direct evidence for harm to human health.[13]

The only comparison I can concoct in which coal waste is not more dangerous than nuclear waste—potentially, hypothetically, actually, or what have you—is on a pound-for-pound basis. But this is clearly not a fair comparison because there are 5 million pounds of coal waste for every pound of nuclear waste generated in producing a given quantity of electricity.

High-Level Radioactive Waste—Hazards and Protective Barriers

Probably the most frequent question I receive about nuclear power is, "What are we going to do with the radioactive waste?" The answer is very simple: we are going to convert it into rocks and put it in the natural habitat for rocks, deep underground. The next question is usually, "How do you know that will be safe?" The answer to that question is the subject of this section and the next.

All of our discussion will be in terms of the high-level waste produced by one large power plant in one year. The waste from such a plant is contained inside the fuel rods, 12-ft-long, $1/2$-in.-diameter cylinders of uranium fuel sealed in metal tubes. One third of these, about 30 tonnes,* are removed from the reactor each year. It will be shipped to a chemical reprocessing plant where it will be dissolved in acid and put through chemical processes to remove 99.5% of the uranium and plutonium which are valuable as fuels for future use. The residue—1.5 tonnes of high-level waste—will then be incorporated into a glass in the form of perhaps 30 cylinders, each about 12 inches in diameter and 10 feet long, weighing about 1000 lb.[15] There is a great deal of research on materials that may be superior to glass as a waste form, but far more is now known about glass, a material whose fabrication is simple and well developed. There is no evidence that it is not satisfactory.[16] Many natural rocks are glasses, and they are as durable as other rocks.†

This 15 tonnes of waste glass, roughly one truckload, will then be shipped to a Federal repository where it will be permanently emplaced deep underground. The U.S. Department of Energy is now spending over $200 million per year in developing this waste storage technology and selecting suitable sites, with this money to be recovered by a tax on nuclear electricity. The estimated cost of handling and storing our one truckload of waste is $5 million,[17] which corresponds to 0.07 cents per kilowatt hour, just over 1% of the cost of electricity to the consumer. I have often been told "We don't want a waste dumping ground in our area." This conjures up a picture of dump trucks driving up and tilting back to allow their loads of waste to slide down into a hole in the ground. Clearly it doesn't cost $5 million for each

---

* 1 tonne = 1 metric ton = 1.1 short tons = 2200 lb.

† From our experience with tumblers, jars, and trinkets, we think of glass as fragile, but a large block of solid glass is not much more fragile than most rocks. Both can crack and chip when forces are applied, but this does not greatly affect their underground behavior since there are powerful pressures holding the pieces together.

truckload to do this, especially on a mass handling basis. What is planned, rather, is a carefully researched and elaborately engineered emplacement deep underground in a mined cavity.

It may be of incidental interest to point out that the present $5 million price tag on storing one year's waste from one plant is several *hundred* times higher than the cost of the original plan as formulated in the 1960s. Those conversant with the earlier plan are still convinced that it provided adequate safety, although the added expenditures do, of course, contribute further safety. This tremendous escalation in the scope of the project is a tribute to the power of public hysteria in a democracy like ours—whether or not that hysteria is justified. It is also an application of Parkinson's Law—it was clearly acceptable to devote one or two percent of the cost of electricity to disposal of the waste, so the money was "available," and being available, the government has found ways to use it.

There have been many safety analyses of waste repositories,[19] and all of them agree that the principal hazard is that somehow the waste will be contacted by ground water, dissolved, and carried by the ground water into wells, rivers, and soil. This would permit contamination of drinking water supplies, or being picked up by plant roots, thereby getting into food. It could thereby get into human stomachs.

The chance of becoming suspended in air as a dust and inhaled by humans is very much less because ground water rarely reaches the surface; moreover, we inhale only about 0.001 grams of dust per day* (and 95% of this is filtered out by hairs in the nose, pharynx, trachea, and bronchi and removed by mucous flow[18]), whereas we eat over 1000 grams of food per day.* On an average, an atom in the top inch of U.S. soil has one chance in 10 billion per year of being suspended as a dust and inhaled by a person, whereas an atom in a river, into which ground water normally would bring the dissolved waste, has one chance in 10 thousand of entering a human stomach. Thus, an atom of waste has a much better chance of becoming absorbed into the body with food or water than with inhaled dust.

External irradiation by radioactive materials in the ground is also a lesser problem. Rock and soil are excellent shielding materials; while the waste remains buried, not a single particle of radiation from it is ever expected to reach the surface*—compare this with the 15,000 particles of radiation from natural sources that strike each of us every second (cf. Chapter 2). If radioactivity is released by ground water, this shielding is still effective as long as it remains underground or dissolved in river water. If the radioactivity does

* Cf. Appendix.

somehow become deposited on the ground surface, it is soon washed away or into the ground by rain. The great majority of the radiation is absorbed by building materials, clothing, or even the air. Thus there would be relatively little exposure to the human body due to radiation originating from materials on the ground.

Quantitative calculations[20] show that food and water intake is the most important mode of exposure from radioactive waste, a position agreed upon in all safety analyses.[19] Hence we will limit our consideration to that pathway. In order to estimate the hazard, we therefore start by determining the cancer risk from eating or drinking the various radioactive materials in the waste.

The method for calculating this is explained in the Appendix, but the results[21] are shown in Fig. 17. For present purposes we need only look at the thick line which is above all the others. This is a plot of the number of

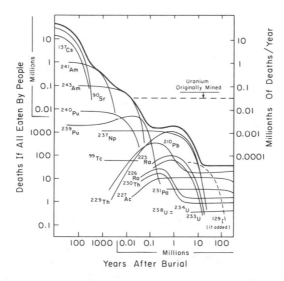

Fig. 17. Toxicity of high-level radioactive waste versus time.[21] The ordinate is the number of cancer deaths that would be expected if all the waste produced by one large nuclear power plant in one year were eaten by people. The individual curves show the toxicity of the individual radioactive species in the waste (as labeled), and the top thick curve shows their sum, the total toxicity. The horizontal dashed curve shows the number of deaths expected if the uranium mined out of the ground to produce the fuel from which this waste was generated were fed to people. The scale on the right shows the deaths per year expected if the buried waste behaves like average rock which is continuously submerged in groundwater.

cancers that would be expected if all of the waste produced by one plant in one year were fed to people.

If this were done shortly after burial, according to Fig. 17 we would expect 50 million cancers, whereas if it were done after 1000 years there would be 300,000 and if after a million years there would be 2,000. Note that we did not specify how many people are involved because that does not matter.*

If all the electricity now used in the United States were generated from nuclear power, we would need about 250 plants, so the waste produced each year would be enough to kill (250 × 50 million =) over 10 billion people. I have authored over 200 scientific papers over the past 35 years presenting tens of thousands of pieces of data, but that "over 10 billion" number is the one most frequently quoted. Rarely quoted, however, are the other numbers given with it[20]—we produce enough chlorine gas each year to kill 400 trillion people, enough phosgene to kill 20 trillion, enough ammonia and hydrogen cyanide to kill 6 trillion with each, enough barium to kill 100 billion, and enough arsenic trioxide to kill 10 billion. All of these numbers are calculated, as for the radioactive waste, on the assumption that *all* gets into people. I hope these comparisons dissolve the fear that, in generating nuclear electricity, we are producing unprecedented quantities of toxic materials.

One immediate application of Fig. 17 is to estimate how much of the waste glass, converted into digestible form, would have a good chance of killing a person who eats it—this may be called a "lethal dose." We calculate it by simply dividing the quantity of waste glass, 16 tonnes, by the values given by the curve in Fig. 1, and the results are as follows:

shortly after burial:	0.01 oz.
after 100 years:	0.1 oz.
after 600 years:	1 oz.
after 20,000 years:	1 lb.

Lethal doses for some common chemicals are as follows:[22]

selenium compounds:	0.01 oz.
potassium cyanide:	0.02 oz.
arsenic trioxide:	0.1 oz.
copper:	0.7 oz.

---

* For example, if feeding a quantity of waste to a thousand people would give each person a risk of 1/10, we would expect (1000 × 1/10 =) 100 deaths; but if instead the same total quantity were fed to 10,000 people, each would eat one-tenth as much and hence have a risk of only 1/100, so we would expect (10,000 × 1/100 =) 100 deaths, the same as before.

Note that the nuclear waste becomes less toxic with time because radioactive materials decay away, leaving a harmless residue, but the chemicals listed retain their toxicity forever.

By comparing these two lists we see that radioactive waste is not infinitely toxic, and in fact it is no more toxic than some chemicals in common use. Arsenic trioxide, for example, used as an herbicide and insecticide, is scattered on the ground in regions where food is grown, or sprayed on fruits and vegetables. It also occurs as a natural mineral in the ground, as do other poisonous chemicals for which a lethal dose is less than one ounce.

Since the waste loses 99% of its toxicity after 600 years, it is often said that our principal concern should be limited to the *short-term,* the first few hundred years. Some people panic over the requirement of security even for hundreds of years. They point out that very few of the structures we build can be counted on to last that long, and that our political, economic, and social structures may be completely revolutionized within that time period. The fallacy in that reasoning is that it refers to our environment here on the surface of the earth where it is certainly true that most things don't last for hundreds of years. However, if you were a rock 2000 feet below the surface, you would find the environment to be very different. If all the rocks under the United States more than 1000 feet deep were to have a newspaper, it couldn't come out more than once in a million years, because there would be no news to report. Rocks at that depth typically last many tens of millions of years without anything eventful occurring. They may on rare occasions be shaken around or even cracked by earthquakes or other diastrophic events but this does not change their position or their interaction with their surroundings.

One way to comprehend the very *long-term* toxicity in the waste is to compare it with the natural radioactivity in the ground. The ground is, and always has been, full of naturally radioactive materials—principally potassium, uranium, and thorium. On a long-term basis (thousands of years or more) burying our radioactive waste would increase the total radioactivity in the top 2000 ft of U.S. rock and soil only by one part in 10 million.[5] Of course the radioactivity is more concentrated in a waste repository, but that doesn't matter. The number of cancers depends only on the total number of radioactive atoms that get into people, and there is no reason why this should be larger if the waste is concentrated than if it is spread out all over the country. In fact quite the opposite is the case: being concentrated in a carefully selected site *deep* underground with the benefit of several engineered safeguards and subject to regular surveillance provides a given atom far less chance of getting into a human stomach than if it is randomly located in the ground. Incidently, it can be shown[5] that the natural radioactivity deep underground is doing virtually no damage to human health, so adding a tiny amount to it is essentially innocuous.

When I point out that concentration of the waste in one place does not increase the health hazard, listeners often comment that it does mean that the risks will be concentrated on the relatively small number of people who live in that area. That, of course, is true, but it is also true for just about any other environmental problem. We in Pittsburgh suffer from the air pollution generated in making steel for the whole country; citizens of Houston and a few other cities bear the brunt of the considerable health hazards from oil refineries that make our gasoline, and there are any number of similar examples. The health burden from these inequities is thousands of times larger than those from living near a nuclear waste repository will ever be.

Figure 17 shows the toxicity of the uranium ore that was originally mined out of the ground to produce the fuel from which the waste was generated. Note that it is larger than the toxicity of the waste after 15,000 years; after that the hazard from the buried waste is less than that from the ore if it had never been mined.* After 100,000 years, there is more radioactive toxicity in the ground directly above the respository (i.e., between the buried waste and the surface) due to the natural radioactivity in the rock than in the repository itself.[5]

A schematic diagram of the waste package, as buried, is shown in Fig. 18. It is enlightening to consider the protections built into the system to mitigate the hazards. For the first few hundred years when the toxicity is rather high, there is a great deal of protection through various time delays before the waste can escape through groundwater. First and foremost, the waste will be buried in a rock formation in which there is little or no groundwater flow, and in which geologists are as certain as possible that there will be little groundwater flow for a very long time. If the geologists are wrong and an appreciable amount of water flow does develop in the rock formation in which the waste is buried, it would first have to dissolve away large fractions of that rock—roughly half of the rock would have to be dissolved before half of the waste is dissolved. This factor might seem to offer minimal protection if the waste is buried in salt, since salt is readily dissolved in water. However, in the New Mexico area being considered for an experimental repository, if all the water now flowing through the ground were diverted to flow through the salt, it would take a million years to dissolve the salt enclosing a repository.[20] The quantity of salt is enormous, while the water flow is not a stream but more like dampness, slowly progressing through the ground.

---

* We do not consider here the residue from uranium mining, which will be discussed in the next chapter.

Fig. 18. Schematic diagram of buried waste package. Components are as follows: Waste glass—the waste itself, converted into a glass. Container—stainless-steel can in which glass is originally cast. Stabilizer—filler material to improve physical and chemical stability of waste. Casing—special material highly resistant to corrosion by intruding water; it should keep water out. Overpack—provides additional corrosion resistance and structural stability. Sleeve—liner for hole, gives structural support. Backfill—material to fill space between waste package and rock, swells when wet to keep water out; if waste becomes dissolved, adsorbs it out of escaping water.

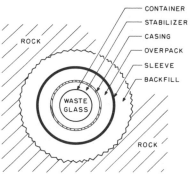

The next layer of protection is the backfill material surrounding the waste package (cf. Fig. 18). This will be a clay which tends to swell up when wet to form a tight seal keeping water flow away from the package.

If water should penetrate the backfill, before it can reach the waste it must get past the metal casing in which the waste glass is to be sealed. Materials for this casing have been developed which give very impressive resistance to corrosion. The present favorite is a titanium alloy which was tested in a very hot (480°F versus a maximum expected repository temperature of 250°F) abnormally corrosive solution, and even under these extreme conditions, corrosion rates were such that water penetration would be delayed for a thousand years.[23] Under more normal groundwater conditions, these casings would retain their integrity for hundreds of thousands of years. Thus, the casings alone provide a rather complete protection system even if everything else fails.

The next layer of protection is the waste form, probably glass, which is not readily dissolved. Glass artifacts from ancient Babylonia have been found in river beds where they have been washed over by flowing river water—not just by the slowly seeping dampness which better describes groundwater conditions—for 3000 years without dissolving away.[24] A Canadian experiment[25] with waste glass buried in soil permeated by ground water indicates that it will last for a hundred million years—i.e., only about 1/100 millionth dissolves each year.

But suppose that somehow some of the waste did become dissolved in groundwater. Groundwater moves very slowly, typically at less than one foot

foot per day—at the New Mexico site it moves only about one inch per day.[26] Furthermore, groundwater deep underground does not ordinarily travel vertically upward toward the surface; it rather follows the rock layers, which are essentially horizontal, and hence it typically must travel about 50 miles before reaching the surface—in the New Mexico area it must travel over a hundred miles.[26] Anyone can easily calculate that to travel 50 miles at one foot per day takes about 1000 years, so this again gives a very substantial protection for the few hundred years that most concern us here. (Long-term effects will be considered later.)

But the radioactive material does not move with the velocity of the groundwater. It is constantly filtered out by adsorption on the rock material, and as a result it travels hundreds or thousands of times more slowly than the groundwater.[27] If the groundwater takes a thousand years to reach the surface, the radioactive materials should take hundreds of thousands or millions of years. Moreover, there are abundant opportunities along the way for waste materials to precipitate out of solution and become a *permanent* part of the rock.[28]

We thus have seven layers of protection preventing the waste from getting out during the first few hundred years when it is highly toxic—(1) the absence of groundwater, (2) the insolubility of the surrounding rock, (3) the sealing action of the backfill material, (4) the corrosion resistance of the metal casing, (5) the insolubility of the waste glass itself, (6) the long time required for groundwater to reach the surface, and (7) the filtering action of the rock. But even if all of these protections should fail, the increased radioactivity in the water would be easily detected by routine monitoring and the water would not be used for drinking or irrgiation of food crops; thus there would still be no damage to human health.

QUANTITATIVE RISK ASSESSMENT FOR HIGH-LEVEL WASTE

Our next task is to develop a *quantitative* estimate of the hazard from buried radioactive waste. Our treatment[28] is based on an analogy between buried radioactive waste and ordinary average rock which is submerged in ground water. We will justify this analogy later, but it is very useful because we know a great deal about the behavior of average rock and its interaction with groundwater. Using this information in a calculation outlined in the Appendix demonstrates that an atom of average rock submerged in groundwater has *one chance in a trillion* each year of being dissolved out of the rock, eventually being carried into a river, and ending up in a human stomach.[28]

In the spirit of our analogy between this average rock and buried radio-active waste, we assume that this probability also applies to the latter; that is, an atom of buried radioactive waste has one chance in a trillion of reaching a human stomach each year. The remainder of our calculation is easy because we know from Fig. 17 the *consequences* of this waste reaching a human stomach. All we need do is multiply the curve in Fig. 17 by one trillionth to obtain the number of fatal cancers expected each year. This is done by simply relabeling the vertical scale as shown on the right side of Fig. 17. For example, a thousand years after burial if all of the waste were to reach human stomachs there would be 300,000 deaths according to the scale on the left side of Fig. 17, but since only one-trillionth of it reaches human stomachs each year, there would be (300,000 ÷ 1 trillion =) 1/3,000,000 deaths per year as indicated by reading the curve with the scale on the right side of Fig. 17. This result is equivalent to one chance in 3 million of a single death each year.

What we really want to know is the total number of people who will eventually die from the waste. Since we know how many will die each year, we simply must add these up. Totalling them over millions of years gives 0.014 eventual deaths.[21] One might ask how far into the future we should carry this addition, and therein lies a complication that requires some explanation.

From the measured rate at which rivers carry dissolved and suspended material into the ocean each year, it is straightforward to calculate that the surfaces of the continents are eroding at an *average rate of* 1 meter (3.3 feet) of depth every 22,000 years.[29] (All the while the continents are being uplifted so that on an average the surface remains at about the same elevation.) Nearly all of this erosion comes from river beds and the minor streams and runoff that feed them. However, rivers change their courses frequently; changing climates develop new rivers and eliminate old ones (e.g., 10,000 years ago the Arizona desert was a rain forest); and land areas rise and fall under geological pressures to change drainage patterns. Therefore averaged over very long time periods, it is reasonable to assume that most land areas erode away at roughly the average rate, one meter of depth every 22,000 years (a result which we will use later in several contexts). This immediately suggests that the ground surface will eventually erode down to the level of the buried radioactive waste. Since the waste will be buried at a depth of about 600 meters, this may be expected to occur after about (600 × 22,000 =) 13 million years. When it happens, the remaining radioactivity will be released into rivers, and according to estimates developed in the Appendix, one part in 10,000 will get into human stomachs.

Therefore, the process of adding up the number of deaths each year should be discontinued after 13 million years; that was done in calculating the number of deaths given above as 0.014. To this we must add the number of deaths caused by release of the remaining radioactivity. A simple calculation given in the Appendix shows that this leads to 0.004 deaths. Adding that number to the 0.014 deaths during the first 13 million years gives 0.018 as the total number of eventual deaths from the waste produced by one power plant in one year.[21] This is the final result of this section, 0.018 eventual fatalities. Recall now that air pollutiron, one of the wastes from coal burning, kills 25 people in generating the same amount of electricity[4]; perhaps new pollution control technology may reduce this number to 5 deaths per year. Even if this comes to pass, we see that the wastes from coal burning are hundreds of times more injurious to human health than are the high-level nuclear wastes.

Another way of understanding the results of our analysis is to consider a situation in which all of the present electricity use in the United States has been provided by nuclear power plants—about 250 would be required—continuously for millions of years. We would then expect (250 × 0.018 =) 4.5 cancer deaths *per year* on an average from all of the accumulated waste. Since, as discussed in Chapter 2, we expect one cancer death for each 8 million millirem of whole-body radiation exposure,[30] the equivalent total radiation exposure to the U.S. public would be (8 million × 4.5 =) 36 million millirem, or 0.15 millirem per person each year. Every American would then have a lifetime radiation exposure of (70 × 0.15 ≅) 10 millirem, giving him a risk of cancer from this radiation of (10/8 million =) 1.3 chances in a million. This is roughly equal to our present risk of being killed in an accident[31] each *day* of our lives.* For the next several thousands of years, the radiation exposures and the risks would be much smaller.

Before closing this section, readers with a scientific bent may be interested in the justification for the basic assumption used in arriving at our estimate, the analogy between buried radioactive waste and average rock which is submerged in flowing groundwater. Actually there are three ways in which the waste may be less secure:

(1) The radioactivity in the waste continues to generate heat for some time after burial, making the buried waste considerably hotter than average rock. There has been some fear that this heat may crack the surrounding rock, thereby introducing easy pathways for groundwater to reach the waste and/

---

* About 110,000 Americans are killed in accidents each year,[31] or 300 per day. Since the U.S. population is 230 million, on an average each of us has a daily risk of (300/230 million =) 1.3 chances in a million.

or to carry off the dissolved radioactivity. This question has been extensively studied since the late 1960s, and the conclusion by now[32] is that rock cracking is not a problem unless the temperature gets up to about 650°F. The actual temperature of the waste can be controlled by the extent to which it is diluted with inert glass and by the intervals chosen for burial of the waste packages. More dilution and larger spacings result in lower temperatures. In the repositories now envisioned these factors will be adjusted to keep the temperature below 250°F, which leaves a large margin of safety relative to the 650°F danger point. If it is decided that this margin should be larger, the easiest procedure would be to delay the burial, since the heat evolving from the waste decreases tenfold after a hundred years, and a hundred-fold after 200 years.[20] Storage facilities for waste packages have been designed and there would be no great difficulty or expense in building them.

(2)The waste glass will be a foreign material in its environment and hence will not be in chemical equilibrium* with the surrounding rock and groundwater. However, chemical equilibrium is a surface phenomenon.[28] When groundwater reaches a foreign material it rapidly dissovles off a microscopically thin surface layer, replacing it with material that had been dissolved out of neighboring rocks. As this surface layer builds up in thickness, it shelters the material beneath from further interaction with the groundwater. This interaction is therefore effectively stopped after only a minute quantity of the foreign material is dissolved, and the foreign material is thereafter no longer "foreign," as its surface is in chemical equilibrium with its surroundings. The fact that the waste glass is a foreign material in its rock environment is therefore not important.

(3) Shafts must be dug from the surface in order to emplace the waste packages, providing access routes for intrusion of water and subsequent escape of dissolved waste that are not available to ordinary rock. The seriousness of this problem depends on how securely the shafts can be sealed. That is a question for experts in sealing, and I am not one of these. However, the general opinion among the experts is that the shafts can be sealed to make them at least as secure as the original undisturbed rock.[33]

Since we have discussed the three ways in which buried waste may be *less* secure than the average rock used in our analogy, it is appropriate to point out ways in which it is *more* secure. In the first place, burial will be in an area free of groundwater flow whereas the rock we considered was already submerged in flowing groundwater. Second, it will be in a carefully

---

* Chemical equilibrium is a situation reached after all rapid reactions between the materials present have taken place.

chosen rock environment, whereas the rock in our analogy is in an *average* environment; it might, for instance, be in an area of extensive rock fracturing. Third, the waste package will be in a special casing providing extreme resistance to water intrusion, a protection not available to average rock; this casing should provide adequate protection even if everything else imaginable goes wrong. And finally, if any radioactive waste should escape and get into food and water supplies, it would very probably be detected by routine monitoring programs in plenty of time to avert any appreciable health consequences.

In balancing the ways in which buried waste is more secure and less secure than the average rock submerged in groundwater with which we compared it, it seems reasonable to conclude that our analogy is a fair one, and if anything it overestimates the dangers from the waste.

It should be pointed out that there are other methods than the one presented here for quantifying the probability for an atom of material in the ground to enter a human stomach. For example,[20] we know the amounts of uranium and radium in the human body from measurements on cadavers, and we know the amounts in the ground. From the ratios of these, we can estimate the probability per year for an atom of these elements in the ground to enter a human body. This gives a result in good agreement with that of the calculation we presented.

Sometimes when I complete a demonstration like the one above of the safety of buried radioactive waste, a listener complains that the demonstration is invalid because we don't know how to bury waste. That, indeed, is a viewpoint often presented in the news media. How anyone can possibly be so naive as to believe that we don't know how to bury a simple package is truly a challenge to my imagination. Many animals bury things, and people have been burying things for millions of years. Any normal person could easily suggest several different ways of burying a waste package. Why, then, are we devoting so much time and effort to developing a waste burial technology? The problem here is to decide on the *best* way of burying the waste to maximize its security. Whether it is worthwhile to go to such pains to optimize the solution is highly questionable, especially since we have shown that even the haphazard burial of average rock by natural processes would provide adequate safety. But the public's irrational worries about the dangers have made our government willing to spend large sums of money on the problem (to be recovered later in our electricity bills). That is why so much time and money are being consumed in determining the very best possible way of burying the radioactive waste.

Long-Term Waste Problems from Chemical Carcinogens

In order to appreciate the meaning of the result obtained in the last section, 0.018 eventual cancer deaths from each plant-year of operation, it is useful to put it in perspective by comparing it with the very long-term cancer risks from the wastes produced by other methods of generating electricity. Here we consider certain types of chemicals that cause cancer; other examples will be given in the next chapter.

For radioactive waste we have shown that average radiation exposures will be a small fraction of one millirem per year, hundreds of times less than our normal exposure to natural radiation. The only reason it has a calculable effect is that we are using a linear dose–response relationship, assuming, for example, that if a million millirem causes a given risk, one millirem will cause a millionth of that risk. This linear relationship does not apply to most toxic chemicals, like carbon monoxide poisoning, for which there is a "threshold" exposure required before there can be any harm. However, it is now widely believed that a linear, no-threshold relationship is valid for chemical carcinogens (i.e., chemicals that can induce cancer) just as for radiation. Basically the reason for this thinking in both cases is that cancer starts from injury to a single molecule on a single chromosome in the nucleus of a single cell, often induced by a single particle of radiation or by a single carcinogenic molecule. The probability for this simple process is not dependent on whether or not other carcinogenic molecules are causing injury elsewhere. Thus the probability of cancer is simply proportional to the number of carcinogenic molecules taken into the body, as described by a linear relationship. This linear, no-threshold dose–response relationship for chemical carcinogens has now been officially accepted by all U.S. Government agenices charged with responsibilities in protecting public health such as the Environmental Protection Agency, the Occupational Safety and Health Administration, and the Food and Drug Administration.[34]

The particular carcinogens of interest in our discussion are the chemical elements cadmium, arsenic, beryllium, nickel, and chromium. There is a considerable body of information indicating that each of them causes cancer, and enough quantitative information to estimate the risk per gram of intake.[35,36] All of them are present in small amounts in coal. Therefore, when coal is burned, they are released, and eventually by one route or another, end up in the ground. Small amounts of each of them are also contained in food. As in the case of all materials in food, they are taken in from the ground through the roots of plants.

As in the case of radioactive waste, the principal problem in evaluating the hazard is to estimate the probability for an atom of these elements in the ground eventually to reach a human stomach. However, the problem is simpler here because the toxicity in these carcinogenic elements does not decay with time as does that of radioactive waste, shown in Fig. 17. Since some quantity of these elements is ingested by humans every year, if an atom remains in the ground indefinitely, eventually after a long enough time, it has a good chance of reaching a human stomach. However, there are other processes for removing materials from the ground, the most important being erosion of soil, with the materials being carried by rivers into the oceans. The probability for an atom in the ground to enter a human stomach is therefore the probability for this to happen before it is washed into the oceans. This is just the ratio of the rate for entering human stomachs to the rate of removal by erosion.

The rates for both of these are known. From chemical analyses of food and the quantity consumed each year, we can calculate the rate at which cadmium, for example, enters human stomachs in the United States.[37] From the rate at which American soil is washed into oceans and the measured abundance of cadmium in soil, we can calculate the rate at which cadmium is carried into oceans.[29] We thus can determine the probability for an atom of cadmium in the soil to enter a human stomach. It turns out to be surprisingly high, about 1.3%. It is five times smaller for arsenic, 20 times lower for beryllium and nickel, and 60 times lower for chromium[35] (cadmium atoms are more easily picked up by plant roots because of their chemical properties).

Since we know the quantities of each of these carcinogenic elements deposited into the top layers of the ground as a result of coal burning,[38] the probability for them to be transferred from the ground into humans, and their cancer risk once ingested, it is straightforward to estimate the number of deaths caused by coal burning. The result[35] is about 50 deaths due to one year's releases from one large plant. These deaths are spread over the time period during which the top few meters of ground will be eroded away, about 100,000 years.*

If one takes a much longer-term viewpoint, these deaths would have eventually occurred even if the coal had not been mined and burned, because after several millions of years, erosion would have brought this coal near to the surface where its cadmium, arsenic beryllium, chromium, and nickel could

---

* We have shown that the average erosion rate is one meter every 22,000 years. We imply here that most of the material ends up within 4.5 meters (15 feet) of the surface (4.5 × 22,000 = 100,000).

have been picked up by plant roots and gotten into food. However, the fact that the coal was mined means[39] that its turn near the surface is taken by other rock which, we assume, has the average amount of these carcinogenic elements in it. This carcinogenic material represents a net additional health risk to humankind due to the use of coal. Taking this into account, the final result[35] is that there would be a net excess of 70 deaths. Note that this is thousands of times larger than the 0.018 deaths from high-level radioactive waste.

The chemical carcinogens also have an important impact on the long-term health consequences of electricity produced by solar energy. We have already pointed out that producing the materials for deployment of a solar array requires about 3% as much coal burning as producing the same amount of electricity by direct coal burning.[40] The quantity of chemical carcinogens released in the former process is therefore 3% of those released in the latter, so in the very long term we may expect about 2 eventual deaths (3% of 70) from an amount of solar electricity equivalent to that produced by a large nuclear or coal-fired power plant in one year.

But there may be much more important problems. A prime candidate material for future solar cells is cadmium sulfide. When these cells deteriorate, that material will probably be disposed of in one way or another into the ground. The cadmium introduced into the ground in this way[41] is estimated to cause, eventually, 80 deaths[35] for the quantity of electricity produced by a large nuclear plant in one year. Most of the cadmium used in this country is imported. Thus, in contrast with the situation for coal, it would not have eventually reached the surface of U.S. soil without this technology. Again we encounter consequences thousands of times higher than the 0.018 deaths from radioactive waste.

Should We Add Up Effects over Millions of Years?

Many may be bothered by the idea of adding up effects over millions of years as we have been doing in this chapter. Personally, I agree with them. The idea of considering effects over such long time periods was introduced by antinuclear theorists, who often insist that it be done. The analyses presented above show that such a procedure turns out very favorably for nuclear power. However, there are many good reasons for confining our attention to the "forseeable future," which is generally interpreted to be 100–1000 years. Several groups, including most government agencies, have adopted 500 years as a reasonable time period for consideration, so we will use it for this discussion.

The reason for ignoring lives that may be lost in the far future is *not* because those lives are less valuable than our own or those of our closer progeny. There surely can be no moral basis for any such claim. But there are at least three reasonable and moral bases for ignoring the distant future.[42]

One such basis is the improvement in cancer-cure rates. If we compare cancers diagnosed between 1960 and 1964 and between 1970 and 1973, the percentage surviving five years after diagnosis[42] increased from 31% to 37% for males, and from 47% to 52% for females; omitting the largest single contributions, lung cancer in males (still largely incurable) and breast cancer in females (about 50% curable), survival improved from 37% to 45% for males, and from 42% to 46% in females. These data roughly correspond to a decrease by about one percent per year in the risk of dying from cancer once it appears. If this rate continues, the risk will be reduced to 81% of its present value after 20 years, 66% after 40 years, 35% after 100 years, and less than 1% after 500 years. Of course it is unlikely that cure rates will improve in such a constant fashion. It is more likely that they will stagnate before long, or on the other hand that there will be a breakthrough leading to a dramatic improvement. But most knowledgeable people agree that there is a good chance that cancer will be a largely curable disease in 500 years. The deaths we calculate to occur after that time will therefore probably never materialize.

The second basis for ignoring deaths calculated to occur in the distant future is what I call the "trust fund" approach. We showed in the last chapter that there are many ways that money can be used to save lives at a rate of at least one life saved per $100,000 spent. But even suppose that so many improvements are instigated that it will cost $1 million to save a life in the distant future (all our discussions discount inflation, so $1 million is somewhat more than an average future person will earn or spend in a lifetime). There is a continuous record extending back 5000 years of money always being able to draw at least 3% real (i.e., discounting inflation) interest.[44] Each dollar invested now at 3% interest is worth $2.5 million after 500 years and will therefore be capable of saving more than one life. It is therefore much more effective in saving far future lives to set up a trust fund to be spent in their time for that purpose rather than for us to spend money now to protect them from our waste. One might wonder about the mechanics and practicality of setting up a trust fund to be transmitted to future generations, but this is no problem. By overspending, we are now building up a public debt that they will have to pay interest on, so rather than actually set up a trust fund, we need only reduce our spending and thereby leave future generations with more money to use for life saving.

A third basis for ignoring the distant future in our considerations is the "biomedical research" approach. If we don't feel comfortable with the "trust fund" approach of putting aside money to be spent in the future for saving lives, we can save future lives by spending money today on biomedical research. One study[45] concludes that the $68 billion spent on such research between 1930 and 1975 is now saving 100,000 lives per year in the United States, one life *per year* for every $680,000 spent. Another study[46] estimates that the $20 billion spent in 1955–1965 is now saving 100,000 lives per year, or one life per year for every $200,000 spent. Age-adjusted mortality rates in the United States have been declining steadily in recent years; if one-fourth of this decline is credited to biomedical research, the latter is now saving one life per year for every $1 million spent.[42] As the easier medical problems are solved and the more difficult ones are attacked, the cost goes up. Let us therefore say that for the near future, it will cost $5 million to save one life *per year* in the United States, not to mention that it will save lives in other countries. Over the next 500 years, saving one life per year will avert a total of 500 deaths, a cost of $10,000 per life saved. If we extend our considerations beyond 500 years, the price drops proportionally. Since we are now spending many millions of dollars per future life saved in handling our radioactive waste, it would clearly be much more beneficial to future generations if we put this money into biomedical research. This situation is already ridiculous, but extending considerations beyond 500 years would make it utterly absurd.

The straightforward way to implement this alternative would be for the nuclear industry to contribute money to biomedical research rather than to improved long-term security of nuclear waste, but even that may not be necessary. The nuclear industry pays money to the government in taxes, and the government spends roughly an equal sum of money to support biomedical research. All that is necessary to implement an equitable program from the viewpoint of future citizens is that 0.1% of this cash flow from the nuclear industry-to-government-to-biomedical-research be redefined as support of biomedical research by the nuclear industry. If it would make people happier, the government tax on the nuclear industry could actually be raised by 0.1% and this money could be directly added to the government's biomedical research budget. No one would notice the difference since it is much less than a typical annual change in these taxes and budgets.

As a result of any one of the above three rationales, or a combination of them, many people, including myself, feel that we should only worry about effects over the next 500 years. For their benefit we now develop estimates of the health effects expected during that time period, from one year's operation of a large power plant:

For the high level waste, we take the effects each year from Fig. 17 using the scale on the right side and add them up over the first 500 years. The result is 0.003 deaths, most of them in the first 100 years. However, this ignores all the time delays and special protections during this early time period that were outlined in the discussion. Geologists feel confident that no groundwater will intrude into the water repository for at least a thousand years, the corrosion-resistant casing is virtually impregnable for several hundred years, it takes groundwater a thousand years to get to the surface, etc. Moreover, since it is quite certain that some sort of surveillance would be maintained during this period, any escaping radioactivity would be detected in plenty of time to avert health problems. These factors reduce the dangers at least 30-fold, to no more than 0.0001 deaths.

The chemical carcinogens released into the ground in coal burning do about one percent of their damage during the first 500 years; we may consequently expect 0.5 deaths in that period.[35] With a 500 years perspective, there is no need to consider the much later effects of the coal reaching the surface by erosion if it had not been mined.

By similar reasoning, the chemical carcinogens from solar electricity will cause 0.015 deaths due to the coal used in making materials for it; if cadmium sulfide solar cells are used, they will cause 0.8 deaths.[35]

WHY THE PUBLIC FEAR?

If the picture I have presented in this chapter is correct, does not omit important aspects of the problem, and is not misleading, it is difficult for me to understand why people worry about the dangers of high-level radioactive waste. This material has been published in the appropriate scientific journals for some time, having been approved by scientist referees and editors to get there. There have been no critiques of it offered. I have presented it at international, national, and regional scientific conferences, and in seminar talks at several major universities, all without receiving substantive criticisms or adverse comments. I therefore know of no evidence that my treatment is not a correct and valid assessment of the problem.

There have been other scientific analyses of the high-level radioactive waste hazards using very different approaches which I belive to be less valid than mine, but they come out with rather similar results.[19] They also find that the health effects will be trivial.

Where, then, is the opposition? What is the basis for the public worry about this problem? There have been a few scientific papers that concoct very special scenarios in which substantial—but still far from catastrophic—harm to public health might result, but they never attempt to estimate the likelihood of their scenarios. All of them are obviously extremely improbable. Note that in my treatment the probabilities of all scenarios are automatically taken into account as long as they apply equally to average rock.

Nevertheless, there can be no question but that the fear abounds. A segment on a network TV morning show opened with a statement to the effect that '[this waste] could possibly contaminate the environment. The result, of course, would be too horrible to contemplate; it could eventually mean the end of the world as we now know it.' I cannot imagine any situation in which this last statement is meaningful short of feeding all of the people in the world directly with the waste. As has been pointed out, there are many other substances that would serve with greater effectiveness if this were done.

Probably the best evidence for this fear is that our government is willing to spend huge sums of money on the problem. The fear is surely there. Perhaps the most important reason for it is that disposal of high-level waste is often referred to as an "unsolved problem." When people say this to me, I ask whether disposal of the waste from coal burning—releasing it as air pollution which is killing about 10,000 Americans each year—is a "solved problem." They usually say "no" but add that air pollution is being reduced. Actually we can hardly hope to reduce it to the point of killing less than 2,000 Americans per year. It is therefore difficult to understand how that converts it into a "solved problem." When I point out that the waste problem is not unsolved, since there are many known solutions, the usual retort is "Then why aren't we burying it now?" There are two answers to that question: (1) we are in the process of choosing the best solution, and (2) we don't have any waste to bury now because we're not doing reprocessing of spent reactor fuel (cf. Chapter 7). This is about as far as the discussion goes, but the real difficulty is that it rarely goes that far. The great majority of people are still hung up on "it's an unsolved problem."

I like to point out that nobody is being injured by the high-level waste, and there is no reason to believe that anyone will be injured by it in the foreseeable future—compare this with the 10,000 deaths per year from the wastes released in coal burning. The usual reply is "how do you know that nobody is being injured?" The answer is that we constantly measure radiation doses in and around nuclear facilities and throughout our environment, and nobody is receiving any such doses from high-level waste. If there is no radiation dose, there can be no harm. I don't believe any scientist would argue this point, but I'm sure most people continue to believe that high-level waste (or the spent fuel from which it is to be derived) actually is doing harm now.

The real difficulty with public understanding of the high-level waste problem is that the scientists' viewpoint is not being transmitted to the public. Transmitting information from the scientific community to the public is in the hands of journalists, who have chosen not to transmit on this question. I'd hate to speculate on their motives, but they are doing great damage to our nation. These are the same journalists who constantly trumpet the claim that the government is suppressing information on nuclear energy. What more despicable suppression of information can there be than refusing to transmit the truth about important questions to the public? It's every bit as effective as lying.

APPENDIX

Probabilities for Entering the Human Body

We inhale about 20 cubic meters (m^3) of air per day, or 7000 m^3 per year.[36] Dust levels in air from materials on the ground becoming suspended[4] are about 35 $\times$ 10^{-6} g/m^3. Thus we inhale (20 $\times$ 35 $\times$ 10^{-6} =) 0.7 $\times$ 10^{-3} grams per day of material from the ground, or 0.25 grams per year.

The area of the United States is about 10^{13} m^2, so the volume of the top inch (0.025 m) of soil is (.025 $\times$ 10^{13} =) 2.5 $\times$ 10^{11} m^3. Since the density of soil is 2 $\times$ 10^6 g/m^3, this soil weighs (2 $\times$ 10^6 $\times$ 2.5 $\times$ 10^{11} =) 5 $\times$ 10^{17} grams. Since each person inhales 0.25 g/yr of this soil, the quantity inhaled by the U.S. population (230 $\times$ 10^6) is (230 $\times$ 10^6 $\times$ 0.25 =) 6 $\times$ 10^7 g/ yr. The probability for any one atom in the top inch of U.S. soil to be inhaled by a human is therefore (6 $\times$ 10^7/5 $\times$ 10^{17} =) 1.2 $\times$ 10^{-10}, a little more than one chance in 10 billion.

The probability for an atom in a river to enter a human is very much larger. The total annual water flow in U.S. rivers is 1.5 $\times$ 10^{15} liters,[29] whereas the total amount ingested by humans is 2.2 liters/person per day[37] $\times$ 365 days/year $\times$ 230 $\times$ 10^6 (population), equals 1.8 $\times$ 10^{11} liters per year. Thus the probability for an atom in a river to be ingested by a human is (1.8 $\times$ 10^{11}/

$1.5 \times 10^{15}) = 1.2 \times 10^{-4}$, or a little more than one chance in 10,000 per year.

Radiation Reaching the Surface from Buried Waste

Rock and soil attenuate gamma rays, the most penetrating radiation from radioactive waste, by more than a factor of 3 per foot.[47] Since the waste will be buried 2000 feet underground, gamma rays from it would be attenuated by a factor of $3^{2000} = 10^{1000}$, an enormously large number. The radioactive waste produced by a million years of all-nuclear power in the United States would eventually emit about 10^{36} gamma rays[20]; thus the probability of even a single gamma ray from it reaching the surface would be about $(10^{36}/10^{1000} =)$ 10^{-964}, an infinitesimally small probability (964 zeros after the decimal point).

Derivation of Fig. 17

For illustrative purposes, let us calculate the number of *liver* cancers expected from people eating the quantity Q of plutonium-239 (^{239}Pu) which is present in the waste.[7] Since the radiation emitted by ^{239}Pu, called *alpha particles,* does not go very far—it can barely get through a thin sheet of paper—this material can cause liver cancer only if it gets into the liver. Experiments indicate[48] that 0.01% of ingested ^{239}Pu gets through the walls of the gastrointestinal tract into the bloodstream, and of this, 45% is deposited in the liver; thus the quantity reaching the liver is ($Q \times 0.0001 \times 0.45 =$) $0.000045Q$. From this we can calculate how many alpha particles strike the liver each second. ^{239}Pu remains in the liver for an average of 40 years[48]; multiplication by the number of seconds in 40 years then gives the total radiation exposure to the liver which may be expressed in millirem. This is then multiplied by the risk of liver cancer per millirem of alpha particle bombardment, which is known from studies of patients exposed for medical purposes,[49] to obtain the number of liver cancers expected if the quantity Q of ^{239}Pu were ingested by people; it comes out to be 1400.

But once ^{239}Pu gets into the bloodstream, it can also get into the bone—45% accumulates there, and it stays for the remainder of life[48]; it therefore can cause bone cancer. A calculation like that outlined above indicates that 700 cases are expected. When other body organs are treated similarly, the total number of cancers expected totals 2300, which is the value of the curve labeled "^{239}Pu" in Fig. 17 in early time periods.

The quantity of ^{239}Pu in the waste does not stay constant, for two reasons.

Every time a particle of radiation is emitted, a ^{239}Pu atom is destroyed, causing the quantity to decrease. But ^{239}Pu is the residue formed when another radioactive atom, ^{243}Am, emits radiation which adds to the quantity. When these two effects are combined, the quantity versus time is as shown by the curve in Fig. 17.

There are many other radioactive species besides ^{239}Pu in the waste; similar curves are calculated for them and shown in Fig. 17. The total number of cancers is then the sum of the number caused by each species, which is obtained by adding the curves at each time. The result is the thick curve above all the others in Fig. 17. That is the curve used in our discussions as the number of cancers expected if all the waste produced by one power plant in one year were fed to people in digestible form.

Probability for an Atom in the Ground to Enter a Human Stomach[28]

Consider a flow of groundwater, called an *aquifer*, along a path through average rock and eventually into a river. There is a great deal of information available on aquifers, like their detailed path through the rocks, the amount of water they carry into rivers each year, and the amounts of various materials dissolved in them.[50] From the latter two pieces of information, we can calculate the quantity of each chemical element carried into the river each year by an average aquifer—i.e., how much iron, how much uranium, how much aluminum, etc.

Where did this iron, uranium, etc. in the groundwater come from? Clearly, this material was dissolved out of the rock. From our knowledge of the path of the aquifer through the rock and the chemical composition of rock, we know the quantity of each of the chemical elements that is contained in the rock traversed by the aquifer. We can therefore calculate the fraction of each element in the rock that is dissolved out and carried into the river each year. For example, a particular aquifer may carry 0.003 lb of uranium into a river each year which it dissolved out of a 50-mile-long path through 200 million tons of rock which contains 1,000,000 pounds of uranium as an impurity. (This is typical of the amount of uranium in ordinary rock.) The fraction of the uranium removed each year is then 0.003/1,000,000 or 3 parts per billion. Similar calculations[28] give 0.3 parts per billion for iron, 20 parts per billion for calcium, 7 parts per billion for potassium, 10 parts per billion for magnesium, etc. To simplify our discussion, let us say that 10 parts per billion of everything is removed each year—this is faster than the actual removal rates for most elements. Incidently, this result implies that for ordinary rock which is submerged in groundwater, only 1% is removed per million years

[(10/1 billion) $\times$ 1 million $=$ 0.01], so it will typically last for 100 million years.

It was shown in the first section of this Appendix that an average molecule of water in a river has one chance in 10,000 of entering a human stomach before flowing into the oceans. For materials dissolved in the water, the probability is somewhat smaller because some of it is removed in drinking water purification processes, but the probability is increased by the fact that some material in rivers finds its way into food and enters human stomachs by that route. These two effects roughly compensate one another. We therefore estimate that material dissolved in rivers has one chance in 10,000 of getting into a human stomach.*

Since an atom of the rock has 10 chances in a billion of reaching a river each year and once in a river has one chance in 10,000 of reaching a human stomach, the over-all probability for an atom of the rock to reach a human stomach is the product of these numbers, [(10/1 billion) $\times$ (1/10,000) $=$] one chance in a trillion per year.[28] That is the value used in our discussion.

In considering effects of erosion, we assumed that all the waste would be released into rivers after 13 million years. According to Fig. 17, if all the waste remaining at that time were to get into human stomachs, about 40 deaths would be expected. But we have just shown that if the material is released into rivers, only one atom in 10,000 reaches a human stomach; we thus expect (40/10,000 $=$) 0.004 deaths. That is the result[14] used in our discussion.

REFERENCE NOTES

1. W. C. Clark, *Carbon Dioxide Review—1982* (Clarendon Press, Oxford, 1982).
2. U.S. Senate Committee on Public Works, "Air Quality and Stationary Source Emission Controls" (1975).
3. E. G. Walther, "A Rating of the Major Air Pollutants and Their Sources by Effects," *Journal of the Air Pollution Control Association,* **22,** 352 (1972).
4. R. Wilson, S. D. Colome, J. D. Spengler, and D. G. Wilson, *Health Effects of Fossil Fuel Burning* (Ballinger Publishing Co., Cambridge, Massachusetts, 1980).
5. Calculation by author, unpublished.
6. B. L. Cohen, "The Role of Radon in Comparisons of Environmental Effects of Nuclear Energy, Coal Burning, and Phosphate Mining," *Health Physics,* **40,** 19 (1981).

---

* As presented here, our discussion ignores those who obtain water from wells rather than from rivers. But by treating the shallow aquifers from which wells derive their water as extensions of rivers, the same result is obtained.

7. A. G. Croff and C. W. Alexander, "Decay Characteristics of Once-Through LWR and LMFBR Spent Fuels, High Level Wastes, and Fuel Assembly Structural Materials Wastes," Oak Ridge National Laboratory Report ORNL/TM-7431 (1980).

8. Reference 4 gives the risk from sulfur dioxide (SO_2) as 3.5×10^{-5}/year for 1 microgram SO_2 per meter^3 of air. An average person inhales 7000 m^3 of air per year so this corresponds to inhaling 7000 μg, or 0.007 g of SO_2. The deaths per gram of SO_2 are then $3.5 \times 10^{-5}/0.007 = 0.005$. An average coal-burning plant produces 3×10^8 g of SO_2 per day; this could then cause ($3 \times 10^8 \times 0.005 =$) 1.5 million deaths if it were all inhaled by people.

9. See Fig. 17 below, which gives the effects from eating all the waste produced in one year. This number must be divided by the number of days per year to obtain the effects from one day.

10. B. L. Cohen, "Ocean Dumping of Radioactive Waste," *Nuclear Technology,* **47,** 163 (1980). Some of the numbers quoted in that paper have been changed due to later data but these are incorporated into the results quoted here.

11. "Air Quality Criteria for Particulate Matter and Sulfur Oxides," U.S. Environmental Protection Agency, February (1981); "Air Quality Criteria for Oxides of Nitrogen," U.S. Environmental Protection Agency Report, June (1980).

12. U.S. Environmental Protection Agency, Health Assessment Document for Polycyclic Organic Matter, Report EPA-600/9-79-008 (1979).

13. B. L. Cohen, "Effects of ICRP-30 and BEIR-III on Hazard Estimates for High Level Radioactive Waste," *Health Physics,* **42,** 133 (1982).

14. Council of Environmental Quality, Environmental Quality-1972, Washington, D.C. (1972).

15. U.S. Department of Energy, Technology for Commercial Radioactive Waste Management, Document DOE/ET-0028 (1979).

16. J. E. Mendel *et al,* "Thermal and Radiation Effects on Borosilicate Glass," Battelle Northwest Lab Document BNWL-SA-5534 (1976).

17. "Environmental Aspects of Commercial Radioactive Waste Management," U.S. Department of Energy Document DOE/ET-0029, May (1979).

18. International Commission on Radiological Protection (ICRP), Task Group on Lung Dynamics, "Deposition and Retention Models for Internal Dosimetry of the Human Respiratory Tract," *Health Physics,* **12,** 173 (1966).

19. These are reviewed by C. M. Koplik, M. F. Kaplan, and B. Ross, "The Safety of Repositories for Highly Radioactive Waste," *Reviews of Modern Physics,* **54,**269 (1982); and Interagency Review Group, Report to the President, U.S. Department of Energy Report TID-29442 (1978).

20. B. L. Cohen, "High Level Waste from Light Water Reactors," *Reviews of Modern Physics,* **49,** 1 (1977); The plans for waste burial have changed somewhat since that time including a threefold dilution of the waste in glass. This is taken into account in the numbers given here.

21. B. L. Cohen, "Effects of Recent Neptunium Studies on High Level Waste Hazard Assessments," *Health Physics* **44,** 567 (1983).

22. H. E. Christensen,"Toxic Substances List," U.S. Department of HEW (1974).
23. J. A. Reuppen, M. A. Molecke, and R. S. Glass, "Titanium Utilization in Long Term Nuclear Waste Storage," Sandia National Lab Report, SAND81-2466 (1981).
24. M. F. Kaplan, "Characterization of Weathered Glass by Analyzing Ancient Artifacts," in *Scientific Basis for Nuclear Waste Management,* C. J. M. Northrup (ed.) (Plenum Press, New York, 1980).
25. W. F. Merritt, Atomic Energy of Canada Limited Report AECL-5317 (1976).
26. H. C. Claibourne and F. Gera, "Potential Containment Failure Mechanisms and Their Consequences at a Radioactive Waste Repository in Bedded Salt in New Mexico," Oak Ridge National Laboratory Report ORNL-TM-4639 (1974).
27. K. J. Schneider and A. M. Platt, "High Level Radioactive Waste Management Alternatives," Battelle Northwest Laboratory Report BNWL-1900 (1974).
28. B. L. Cohen, "Analysis, Critique, and Re-evaluation of High Level Waste Repository Water Intrusion Scenario Studies," *Nuclear Technology,* **48,** 63(1980).
29. R. M. Garrels and F. T. McKenzie, *Evolution of Sedimentary Rocks* (Norton Co., New York, 1971).
30. Cf. Chapter 2.
31. National Safety Council, "Accident Facts—1976," Chicago (1976).
32. P. A. Witherspoon, N. G. W. Cook, and J. E. Gale, "Progress with Field Investigations at Stripa," Lawrence Berkeley Lab Report LBL-10559 (1980).
33. Office of Nuclear Waste Isolation (ONWI), "The Status of Borehole Plugging and Shaft Sealing for Geologic Isolation of Radioactive Waste," ONWI-15, Battelle Memorial Inst., Columbus, Ohio (1979); Office of Nuclear Waste Isolation, Repository Sealing Field Testing Workshop, ONWI-239 (1980); D. M. Roy, M. W. Grutzeck, and P. H. Licastro, Evalution of Cement Borehold Plug Longevity, ONWI-30 (1979).
34. U.S. Consumer Product Safety Commission, U.S. Environmental Protection Agency, U.S. Dept. of Health, Education, and Welfare, U.S. Food and Drug Administration, and U.S. Dept. of Agriculture, "Scientific Bases for Identification of Potential Carcinogens and Estimation of Risks," *Federal Register,* **44,** 39858 (1979). See page 39873. See also *Federal Register,* **40,** 28242; **41,** 21402; **43,** 25658; U.S. Environmental Protection Agency, "National Emission Standards for Identifying, Assessing, and Regulating Airborne Substances Posing a Risk of Cancer," *Federal Register,* **44,** 58642, 1979 (see page 58649); U.S. Occupational Safety and Health Administration, "Identification, Classification and Regulation of Potential Occupational Carcinogens," *Federal Register,* **45,** 5002 (1980).
35. B. L. Cohen, "Long Term Consequences of the Linear No Threshold Dose–Response Relationship for Chemical Carcinogens," *Risk Analysis,* **1,** 4, 267 (1982).
36. Carcinogen Assessment Group, U.S. Environmental Protection Agency, "Preliminary Risk Assessment on Beryllium" (1979); Carcinogen Assessment Group, "Summary and Conclusions Regarding Carcinogenicity of Chromium" (1979); Carcinogen Assessment Group, "Preliminary Risk Assessment on Nickel" (1979); Carcinogen Assessment Group, "Final Risk Assessment on Arsenic" (1980); Car-

cinogen Assessment Group, "Assessment of Carcinogenic Risk from Population Exposure to Cadmium in the Ambient Air" (1978).

37. International Commission on Radiological Protection (ICRP), "Report of the Task Group on Reference Man," ICRP Publication No. 23 (1975).

38. J. Cavallaro, A. W. Deurbrouck, H. Shultz, and G. A. Gibbon, "Washability and Analytical Evalution of Potential Pollution from Trace Elements in Coal," Environmental Protection Agency Report EPA-600/7-78-038, March (1978); H. J. Gluskoter, "Trace Elements in Coal; Occurrence and Distribution," EPA Report EPA-600/7-77-064 (1977); C. T. Ford, "Evaluation of the Effect of Coal Cleaning on Fugitive Elements," Bituminous Coal Research Lab Draft Report L-1082 (1980).

39. B. L. Cohen, "Health Effects of Radon from Coal Burning," *Health Physics*, **42**, 725 (1982).

40. P. Masser, "Low Cost Structure for Photovoltaic Arrays," Sandia Laboratory Report, SAND-79-7006 (1979).

41. P. D. Moskowitz *et al.*, "Photovoltaic Energy Technologies: Health and Environmental Effects Document," Brookhaven National Laboratory Report BNL-51284 (1980).

42. B. L. Cohen, "Discounting in Assessment of Future Radiation Effects," *Health Physics* (in press).

43. L. M. Axtell, S. J. Cutler, and M. H. Myers, "End Results in Cancer," NIH Publication 73-272 (1973).

44. S. Homer, *The History of Interest Rates* (Rutgers University Press, New Brunswick, New Jersey, 1977).

45. S. J. Mushkin, *Biomedical Research: Costs and Benefits* (Ballinger Publ. Co., New York, 1979).

46. R. Auster, I. Leveson, and D. Sarachek, "The Production of Health," *Journal of Human Resources* **4**, 411 (1969).

47. A. Brodsky, *Handbook of Radiation Measurement and Protection*, Vol. 1 (CRC Press, Boca Raton, Fla. 1978).

48. ICRP (International Commission on Radiological Protection), "Limits for Intakes of Radionuclides by Workers," ICRP Publication 30 (1979).

49. National Academy of Sciences Committee on Biological Effects of Ionizing Radiation, "The Effects on Populations of Exposure to Low Levels of Ionizing Rsdiation," Washington, D.C. (1980).

50. D. E. White, J. D. Hein, and G. A. Waring, "Chemical Composition of Subsurface Waters," Chap. F in "Data of Geochemistry," U.S. Geological Survey (1963).

Chapter 6 / MORE ON RADIOACTIVE WASTE

While the high-level waste discussed in Chapter 5 is the problem that has received the lion's share of the attention, it is not the only waste problem involved with nuclear power. In fact it is not even the most important one. Several other nuclear waste issues will be considered in this chapter, followed by a summation of the problem as a whole.

In terms of numbers of deaths, the most important waste problems arise from the radioactive gas radon. We begin with its story.

RADON PROBLEMS[1]

Uranium is a naturally occurring element, present in small quantities in all rock and soil, and in materials derived from them such as brick, plaster, and cement. It is best known as a fuel for nuclear reactors but quite aside from the properties that make it useful for that purpose, it is naturally *radioactive,* which means that it decays into other elements, shooting out high-speed particles of radiation in the process. The residual atoms left following its decay are also radioactive, as is the residue from decay of the latter, and so on until the chain is terminated after 14 successive radioactive decays. One step in this chain of decays is *radon*, which is of very special interest because it is *gas*. An atom of radon, behaving as a gas, has a tendency to move away from the location where it was formed and percolate up out of the ground into the air.* Since uranium is in the ground everywhere, atoms of radon—and radioactive elements into which it decays—float around in the air everywhere and are thus constantly inhaled by everyone. Being a gas, radon is very rapidly exhaled. But the elements into which it decays are not gases; hence they tend to stick to the surfaces of our respiratory passages, exposing the latter to radiation, thereby inducing lung cancer.

A great deal is known about this problem because of several situations in which miners were exposed to very high levels of radon in poorly ventilated mines.[2] In uranium mines, especially, there is often an unusually high concentration of uranium in the surrounding rock. The radon evolving from it percolates out into the mine where it remains for some time because its escape routes are limited. The best study of this problem involved a group of 4146 uranium miners who worked in the Colorado plateau region between 1945 and 1969; there were 159 deaths from lung cancer among them up to 1974, whereas only 25 would have been expected from an average similar group of American men.[3] When this situation was recognized in the late 1960s, drastic improvements were introduced in the ventilation, reducing radon levels about 20-fold, which makes present radon exposure one of the less important of the many significant risks in mining.[4]

From the studies of lung cancer among miners, we can estimate the risks in various other radon exposure situations. The most important of these is in the normal environmental exposure we all receive, especially in our homes, which are often poorly ventilated. Radon percolates up from the ground, usually entering through cracks in the floor or along pipe entries; some also comes from plaster and other building materials that contain small amounts of uranium. In some areas it enters homes with water drawn from underground

---

* A gas is not confined by gravity (e.g., it will not stay in an open bottle) and hence can percolate up out of the soil. A solid or a liquid is confined by gravity and will not percolate upward unless pushed by an external force. In Chapter 5 we discussed waste being dissolved in ground water, and the latter moving to the surface. This motion is due to its being pushed by water at higher elevations that is being pulled downward by gravity into the soil. This is the same reason that water flows in pipes from the basement to upper stories of a house, pushed by water from a reservoir at a high elevation that is being pulled downward by gravity. Radioactive waste converted to glass cannot be converted into a gas. It was heated to 1900°F in the conversion to glass; since it did not vaporize then, it will not vaporize later.

sources. By its being trapped inside homes, radon levels indoors are several times higher than outdoor levels.[5,6]

Using the data on miners and applying the linear hypothesis leads to an estimate that this environmental radon exposure is now causing about 10,000 fatal lung cancers each year in the United States,[7] several times more fatalities than are caused by all other natural radiation combined.

We have already mentioned the fact that reducing air leakage from houses to conserve fuel increases the radon level; if this were done in all houses in accordance with government recommendations, many thousands of additional fatalities each year would result.[7,8] Recalling that most scientists estimate the radiation consequences of a full nuclear power program as a few tens of fatalities per year (even the antinuclear activists estimate only 600 fatalities per year) we see that conservation is a far more dangerous energy strategy than nuclear power from the standpoint of radiation exposure!

These environmental radon problems, of course, have no direct connection with nuclear power, but there are connections in other contexts. The most widely publicized of these is the problem of uranium mill tailings.[9] When uranium ore is mined it is taken to a nearby mill and put through chemical processes to remove and purify the uranium. The residues, called the *tailings,* are dissolved or suspended in water which is pumped into ponds. These ponds eventually dry out, leaving what are ostensibly piles of sand. However, they contain the radioactive products from the decay of uranium— only the uranium itself has been removed. The most important of these decay products is thorium-230, which has an average lifetime of 110,000 years and decays into a chain of products includng radon. In fact, most of the radon generated over the next 100,000 years will come from uranium that has already decayed into thorium-230, so removing the uranium has little effect on radon emissions over this period. But a very important effect of mining the uranium is that the source of this radon was removed from underground where the ore was originally situated, to the surface where the mill tailings are located. This allows far more of it to percolate into the air to cause health problems.

A quantitative calculation of this effect indicates that the mill tailings produced in providing a one-year fuel supply for one large nuclear power plant will cause 0.003 fatalities per year with its radon emissions.[10] This may not seem large, but it will continue for about 100,000 years, bringing the eventual fatality toll to something like 300! Fortunately, there are things that can be, and are being, done about this problem. If the tailings piles are covered with several feet of soil (or perhaps other materials), nearly all of the radon decays away during the time it takes to percolate up through the cover—the average lifetime of a radon atom is only 5.5 days. The law[11] now requires that radon emissions from the tailings be reduced 200-fold by covering them with soil (or with other materials). This lowers the eventual fatality toll down from 300 to 1.5. Even this is nearly a hundred times higher than the 0.018 fatality toll from the high-level waste. Anyone who worries about the effects of radioactive waste from the nuclear industry should therefore worry much more about the uranium mill tailings than about the high-level waste. However, for some reason the latter has received the great majority of the media publicity and hence the public concern. It is these that determine the response of our government. The government research program on controlling mill tailings is only a few percent of the size of that on high-level waste. The dollars go to what the media publicize, not to where the danger is.

By far the most important aspect of radon from uranium mining is yet to be discussed. When uranium is mined out of the ground to make nuclear fuel, it is no longer there as a source of radon emission. This is a point which has not been recognized until recently because the radon that percolates out of the ground originates largely within one meter of the surface; anything coming from much further down will decay away before reaching the surface. Since the great majority of uranium mined comes from depths well below one meter, the radon emanating from it was always viewed as harmless. The fallacy of this reasoning is that it ignores erosion. As the ground erodes away at a rate of one meter every 22,000 years, any uranium in it will eventually approach the surface, spending its 22,000 years in the top meter where it does so much damage.[12] The magnitude of this damage is calculated in the Appendix, where it is shown that mining the uranium to fuel one large nuclear power plant for one year will eventually *save* 500 lives. This completely overshadows all other health impacts of the nuclear industry, making it one of the greatest life-saving enterprises of all time if one adopts a very long-term viewpoint.

Before we can count these lives as permanently saved, we must specify what is to be done with the uranium, for only a tiny fraction of it is burned in today's nuclear reactors. There are two dispositions that would be completely satisfactory from the life-saving viewpoint. The preferable one would be to burn it in breeder reactors and thereby derive energy from it. An easy alternative, however, is to dump it into the ocean. Uranium remains in the ocean only for about 3,000,000 years[13] before settling permanently into the bottom sediments. All uranium in the ground is destined anyway to be carried eventually by rivers into the ocean and spend its 3,000,000 years therein. From a long-range viewpoint it makes little difference to human health if it spends that time now or a few million years in the future. But by preventing it from having its 22,000 year interlude within one meter of the ground surface, we are saving numerous lives from being lost due to radon!

If one adopts the position that only effects over the next 500 years are relevant, there is still an important effect from uranium mining because about half of all uranium is surface mined. About 1% of this comes from within one meter of the ground surface where it is *now* serving as a source of radon exposure. It is shown in the Appendix that eliminating this source of exposure by mining will *save* 0.07 lives over the next 500 years, but in the meantime, radon escaping from the covered tailings will cause 0.01 deaths. Thus, the net effect of the radon from uranium mining is to save 0.06 lives as a result of one year's operation of one large nuclear power plant.

We still have one more radon problem to discuss, namely, the radon released in coal burning. Coal contains small quantities of uranium and when the coal is burned, by one route or another, it ends up somewhere in the ground. Again the problem is complicated by the fact that, if the coal had not been mined, erosion would eventually have brought the coal with its uranium to the surface anyhow. The final result,[14] derived in the Appendix, is that the extra radon emissions caused by the burning of coal in one large power plant for one year will eventually cause 30 fatal lung cancers. This toll, like so many others we have encountered here, is thousands of times larger than the 0.018 deaths caused by the high-level waste produced in generating the same amount of electricity from nuclear fuel.

Solar electricity uses a great deal of coal in manufacturing the materials for its deployment[15]—steel, concrete, glass, and aluminum. In fact these use 3% as much coal as would be needed to produce the same amount of electricity by direct coal burning. Solar electricity should therefore be charged with (3% of 30=) one death per year from radon. This again, is far larger than the 0.018 deaths per year from high-level radioactive waste.

ROUTINE EMISSIONS OF RADIOACTIVITY

Nuclear power plants, and more important, fuel reprocessing plants, routinely emit small quantities of radioactive material into the air and into nearby rivers, lakes, or oceans.[5] Antinuclear propaganda made a rather big issue of this[16] in the early 1970s, but there has been relatively little publicity about it in recent years. The quantities of radioactivity released are limited by several Nuclear Regulatory Commission requirements, including one that equipment must be installed to reduce these emissions if doing so will result in the saving of one life from radiation exposure for every $8 million spent.[17] Since new and improved control technologies are steadily being developed, these emissions have been reduced considerably over the past decade, and that trend is continuing. There are also NRC requirements on maximum exposure to any individual member of the public, limiting it to a few percent of his exposure to natural radiation; one percent is typical for those living very close to a nuclear plant. This gives them a risk equivalent to that of driving 4 extra miles per year, or of crossing a street one extra time every three months.

The principal present emissions of importance from the health standpoint are radioactive isotopes of krypton and xenon (Kr–Xe), chemically inert gases which are therefore very difficult to remove from the air; tritium (T), a radioactive form of hydrogen, that becomes an inseparable part of water, used in such large quantities that it would be impractical to store it for long times; and carbon-14 (C-14), which becomes an inseparable part of the omnipresent gas carbon dioxide. Of these, only C-14 lasts long enough to irradiate future generations. According to a recent study by the United Nations Scientific Committee on Effects of Atomic Radiation,[5] the releases associated with one year of operation of one power plant can be expected to cause 0.036 deaths from Kr–Xe, 0.022 from T, and 0.045 from C-14 over the next 500 years and 0.56 over tens of thousands of years. This gives a total of 0.1 deaths over the next 500 years and 0.6 deaths eventually. There are improved technologies for reducing these emissions 2- to 4-fold, some of which are already scheduled to become legally required in the near future.[18] Note again that these effects are many times larger than the 0.0001 deaths during the first 500 years and the 0.018 eventual deaths from high-level waste. They are also larger than the 0.02 deaths estimated by the Rasmussen Report as the average toll from reactor accidents. Here again, we see that public concern is driven by media attention and bears no relationship to actual dangers.

LOW-LEVEL WASTE

In order to minimize the releases of radioactivity into the environment discussed in the last section, there is a great deal of equipment in nuclear plants for removing radioactive material from air and water by trapping it in various types of filters. These filters, including the material they have collected, are the principal component of what is called *low-level waste*, the disposal of which we will be discussing in this section. Other components are things contaminated by contact with radioactive material* like rags, mops, gloves, lab equipment, instruments, pipes, and valves; and various items that were made radioactive by being in or very near the *reactor where they* are exposed to neutrons.† In addition there are wastes from hospitals, research laboratories, industrial users of radioactive materials, etc. which make up about 25% of the total. In general the concentration of radioactivity in low-level waste is a million times lower than in high-level waste—that is the reason for the name—but the quantities in cubic feet of the former are thousands of times larger. It is therefore neither necessary nor practical to provide the low-level waste with the same security as the high level. The former is buried in shallow trenches about 20 feet deep in commercially operated burial grounds licensed by the Federal or State governments.

Since its potential for doing harm is relatively slight, until recently this low-level waste was handled somewhat haphazardly. There was little standardization in packaging, in handling, or in stacking packages in trenches, and little care in covering them with dirt. As a result, trenches often filled with rain water percolating down through the soil which then dissolved small amounts of the radioactivity. The caretakers regularly pumped this water out of the trenches and filtered the radioactivity out of it, allowing no radioactivity to escape.

The situation was radically changed by two very innocuous but highly publicized incidents during the 1970s. In an eastern Kentucky burial ground, a place called Maxey Flats, tiny amounts of radioactivity were found off-site;[19] the quantities were so small that no one could have received as much as 0.1 mrem of radiation[20] (one chance in 80 million of getting a cancer). It is not clear whether the radioactivity had leaked through the ground or was simply a product of sloppy surface operations,[21] but the publicity was enormous at the time. The January 18, 1976 issue of the *Washington Star* carried a story headlined "Nuclear Waste Won't Stay Buried," which began with "Radioactive wastes are contaminating the nation's air, land, and water." When the head of the U.S. Energy Research and Development Agency (predecessor of Department of Energy) testified for his Agency's annual budget, the first 25 minutes of questions were about Maxey Flats, and a similar pattern was followed with the Chairman of the Nuclear Regulatory Commission.

---

* i.e., particles of radioactive material may stick to them.

† Neutrons can induce nuclear reactions which make things radioactive. Alpha particles, beta rays, and gamma rays emitted by radioactive substances ordinarily cannot. There are essentially no neutrons except while the reactor is operating. Thick shield walls prevent neutrons from getting more than a few feet away from the reactor, preventing them from injuring people.

After another Congressional Committee was briefed about the problem,[22] its Chairman stated publicly that it was "the problem of the century." I pointed out to him that due to the high uranium content of the granite in the Congressional Office Building, his staff was being exposed to more excess radiation every day than anyone had received in toto from the Kentucky incident. He made no further alarmist statements, but as a result of all the attendant publicity the burial ground was closed.

The other incident occurred in a western New York State burial ground where there was a requirement that permission be granted from a State agency to pump water out of the trenches. In one instance, this permission was somehow delayed in spite of urgent warnings from the site operators. As a result, some slightly contaminated water overflowed, with completely negligible health consequences (the largest doses were 0.0003 mrem),[23] leading to widespread adverse publicity and closing of the burial ground.

The journalistic coverage of these incidents was unbelievably shoddy. One TV special showed a local woman saying that the color of the water in the creek had changed, but did not point out that any scientist would agree that this could not possibly be due to the radioactivity; moreover the producer was told that the coloring was caused by bulldozing operations nearby.[24] The same program showed a local farmer complaining that his cattle were sick—"Hair's been turning gray, grittin' their teeth, and they're a-dying, going up and down in milk"; there was no mention of the facts that a veterinarian had later diagnosed the problem as a copper and phosphorus deficiency, the cattle had been treated for this deficiency and had recovered, and that the TV crew had been informed about this long before the program was aired.[24] A large portion of a newspaper feature story[25] was devoted to the story of a woman telling about how she was dying of cancer due to radiation caused by the leakage; a 0.1 mrem radiation dose has one chance in 80 million of inducing a cancer, whereas one person in five dies from cancers due to other causes, so there is no more than one chance in 16 million that her cancer was connected with the leakage. Actually it is surely less than that, because her cancer was not of a type normally induced by radiation. Moreover, 0.1 mrem was the maximum possible dose to any person; the dose received by this particular woman was probably very much lower.

But the biggest sin in the journalistic coverage was in lack of perspective. 0.1-mrem exposure, the largest possibly received by any citizen from these incidents, represents a loss of life expectancy of 0.12 minutes, or 7 seconds. In accordance with the discussion in Chapter 4, this is less than one-third the risk of an average street crossing, something we all do several times a day. Why is such a trivial risk deserving of such coverage?

As a result of this publicity and the public's extreme sensitivity to even the slightest radiation exposure from the nuclear industry, plus some honest desire of government officials to improve security, a new set of regulations was recently formulated.[26] They require[27] that (1) the trench bottoms be well above any accumulating groundwater (i.e., the "water table") so as to exclude the possibility of the trenches filling with water; (2) surface covers be installed to minimize water passing through the trenches; (3) the waste be packed in such a way that the package maintains its size and shape even under heavy external pressures, and when wet, or when subject to other potential adverse chemical and biological conditions; (4) packaging material be more substantial than cardboard or fiberboard; (5) there be essentially no liquid in the waste; (6) void spaces between waste packages be carefully filled, and so on.

Since the movement of water through the trenches can only be downward toward the water table, any radioactivity that escapes from the packages can only move downward until it reaches the water table,[28] after which it can flow horizontally with the ground water flow which normally discharges into a river. This movement, as was explained in our discussion of high level waste, must take many hundreds or thousands of years, because the water moves very slowly, and the radioactive materials are constantly being filtered out by the rock and soil.

In order to estimate the hazards from a low-level waste burial ground,[29] we consider two possible routes from the ground into the human stomach: (1) its being picked up by plant roots and thereby getting into food, and (2) its being carried into a river and thereby getting into water supplies.* To evaluate the food pathway, we assume the unfavorable situation in which all of the waste escapes from its packaging and somehow becomes randomly distributed through the soil between the surface and the top of the water table. For natural materials, we know how much of each chemical element there is in this soil[30] and how much there is in food,[31] so we can calculate the probability per year for transfer from the soil into food. This probability then can be applied to the radioactive material of the same element, and of other elements chemically similar to it. When this probability is multiplied by the number of cancers that would result if *all* the low-level waste were ingested by people, calculated using the methods described in the Appendix of Chapter 5 for the quantities of low-level waste generated by the nuclear industry,[32] the product gives the number of deaths per year expected via the food pathway.

It is estimated that it would take something like 800 years for any of the

---

* It could also become suspended in air as a fine dust and inhaled by people, but calculations indicate that this pathway is much less important than those discussed here (see Chapter 5).

radioactivity to reach a river.[28] In order to estimate conservatively the hazards from the water pathway, we therefore assume that *all* radioactivity remaining after 800 years reaches a river,* and as in the case of high-level waste explained in Chapter 5, one part in 10,000 of it enters a human stomach.

Adding the effects of the food and drinking water pathways, we obtain the total number of deaths from the low-level waste generated by one large nuclear power plant in one year to be 0.0001 over the first 500 years, and 0.0005 eventually.

A great deal of detail has been omitted from this discussion, so interested readers might consult the research paper on which it is based.[29]

TRANSURANIC WASTE

Elements heavier than uranium, called "transuranics" (TRU), are not found in nature but are produced in reactors and therefore become part of the waste, both high level and low level. These elements like plutonium, americium, and neptunium, have special properties that make them more dangerous than most other radioactive materials if they get into the body. Since they are often relatively easily separated from other low-level wastes, it is government policy to do this where practical, and to dispose of these TRU wastes by deep burial somewhat similar to that planned for high-level waste. The first repository for this purpose is now under construction in New Mexico. The amount of radioactivity in the TRU waste is many times less than in the high-level waste during most time periods. Furthermore, elevated temperatures which many consider to be the principal threat to waste security are not a factor for TRU waste. Therefore, while there has been little study of commercial TRU waste disposal (there will be no TRU waste until reprocessing is instituted), it seems clear that it represents less of a health hazard than high-level waste.

SUMMARY OF RESULTS

The health effects from electricity generation that have been discussed in this and the previous chapter are listed in Table 1 in terms of the number of deaths caused by one year's operation of a large power plant during the

---

* This was *not* assumed in calculating effects of the food pathway. In those calculations, the radioactive material was assumed to remain in the soil indefinitely.

TABLE 1

SUMMARY OF THE NUMBER OF DEATHS CAUSED BY THE WASTE
GENERATED BY ONE LARGE POWER PLANT IN ONE YEAR, OR BY THE
EQUIVALENT AMOUNT OF ELECTRICAL ENERGY PRODUCTION

Source	Deaths caused	
	First 500 years	Eventually
Nuclear		
High-level waste	0.0001	0.018
Radon emissions	−0.06	−500
Routine emissions (Kr,T,C-14)	0.05	0.3
Low-level waste	0.0001	0.0005
Coal		
Air pollution	5	5
Radon emissions	0.11	30
Chemical carcinogens	0.5	70
Photovoltaics for solar energy		
Coal for materials	0.9	3
Cadmium sulfide	0.8	80

first 500 years, and eventually over multimillion year time periods. The minus signs for radon emissions from nuclear power indicate that lives are *saved* rather than lost.

This table is the "bottom line" on the waste issue. It shows that, in quantitative terms, radioactive waste from nuclear power is very much less of a hazard than the chemical wastes, or even the *radioactive* wastes from coal burning or solar energy. Almost every technology-based industry uses energy derived from coal and produces chemical wastes, and in nearly all cases, these are more harmful than the nuclear waste. This is true ignoring the lives saved by mining uranium out of the ground; if the latter is included, nuclear waste considerations give a tremendous net *saving* of lives. By any standard of quantitative risk evaluation, the hazards from nuclear waste are not anything to worry about.

The problem is that no one seems to pay attention to quantitative risk evaluation. There have been several books written (always by nontechnical authors) about the hazards of nuclear waste without a mention of what the hazards are in quantitative terms. In some cases the authors interviewed me, at which time I went to great lengths in trying to explain this point, but they didn't seem to understand. When I explained what would happen if all the radioactive waste were ingested by people, they were busily taking notes, but when I tried to explain how small the probability is for an atom of buried waste to find its way into a human stomach, the note-taking stopped and they showed impatience and eagerness to get on to other subjects. My impression was that they were writing a book to tell people how horrible nuclear waste is, and that was all they were interested in.

I was once interviewed for a network TV special on nuclear waste, and that was even worse. The production team made it completely obvious that they knew what they wanted to tell the public; what they wanted from me was material to help them in that endeavor, or at least something that would suggest that they were fair in consulting people on both sides of the issue. One thing they didn't want was analyses like those presented here, showing quantitatively how small the risks are. Another thing they didn't want was comparisons between the risks of the waste from coal burning and nuclear energy. They had a point they wanted to make and, by God, they were going to make it—no scientist was going to deter them with facts and calculations. Of course their mission is to attract audiences, and stories about dangers are much better for that purpose than scientific explanations for why the dangers are negligible.

THE REAL WASTE PROBLEM

This behavior by the media and even by authors of books has led to a *real* waste problem—waste of tax-payers' money spent to protect us from the imagined dangers of nuclear waste. One example is the handling of high-level waste from production of materials for nuclear bombs at the Savannah River plant in South Carolina. This military high-level waste is considerably less radioactive than the commercial waste discussed in Chapter 5, so several alternative plans were developed and carefully studied.[33] One relatively cheap ($500 million) one was to pump the waste in solution deep underground, whereas a much more expensive ($2.7 billion) one was to handle it like civilian high-level waste, converting it to glass and placing it in an engineered deep underground repository. It was estimated[33] that the first plan would eventually lead to 8 fatal cancers spread over many thousands of years into the future (assuming that there will be no progress in fighting cancer). The U.S. Department of Energy therefore decided to spend the extra $2.2 billion to save (?) the 8 future lives, a cost of $270 million per life saved. If this $2.2 billion were spent on cancer screening or highway safety, it could save over ten thousand American lives in our generation. As one example, we could put air bags in all new cars manufactured this year and thereby save over a thousand lives per year for the next several years. Or we could install smoke alarms in every American home, saving 2000 lives each year for some time to come.

But even these comparisons do not fully illustrate the absurdity of our unbalanced spending, because people living many thousands of years in the future have no closer relationship to us than people living in India and other underdeveloped countries today where millions of lives could be saved at a cost of $5000 each by simply sending food. This $2.2 billion could save nearly a half million lives there instead of *possibly* saving 8 far future American lives. One could argue that it doesn't really help to save lives in countries suffering from *chronic* starvation because doing so increases their population and thus aggravates the problem, but that argument does not apply to areas where catastrophic one-year famines have struck. For example, during the 1970s, there were two such famines in Bangladesh that killed 427,000 and 330,000, a localized famine in India that killed 829,000, and one each in the African Sahel and Ethiopia that were responsible for 100,000 and 200,000 deaths, respectively.[34] If we place such a low value on the lives of these peoples, how can we place such a high value on the lives of whoever happens to be living in this area many thousands of years in the future? There is little reason to believe that they will be our direct descendants.

West Valley—The Ultimate Waste Problem

However, the most flagrant waste of taxpayer dollars in the name of nuclear waste management is going on at West Valley, New York, about 30 miles south of Buffalo. Since the West Valley problem has been widely publicized, it is worth describing in some detail.[35] This was the site of the first commercial fuel reprocessing plant, completed in 1966 and operated until 1972 when it was shut down for enlargement to increase its capacity. During the following few years, government safety requirements were substantially escalated making the project uneconomical: the original cost of the plant was $32 million, and the initial estimated cost of the enlargement was $15 million, but it would have cost $600 million to meet the new requirements for protection against earthquakes. (All areas have some susceptibility to earthquakes, but it is minimal in the West Valley area.) It was therefore decided to abandon the operation, raising the question of what to do with the high-level waste stored in an underground tank.

The potential hazard was that the radioactive material might somehow leak out, get into the groundwater, and be carried with it into a nearby creek which runs into Lake Erie. Lake Erie drains into Lake Ontario and eventually into the St. Lawrence river, and the three of these are used as water supplies for millions of people. How dangerous would this be?

If all of the radioactive waste stored at West Valley were dissolved in Lake Erie now and if it passed unhindered through the filters of water supply systems with no precautions being taken, we could expect[36] 40,000 eventual fatalities to result. However, the radioactivity decreases with time by about a factor of ten per century for the first few hundred years, so that if it were dumped into Lake Erie 400 years from now, only six fatalities would result; and if the dumping occurred more than 1,000 years in the future, there would probably not be a single fatality.

How likely would it be for wastes to get into Lake Erie in the near future? Let us suppose that all the containment features designed into the system were to fail, releasing all of the radioactive material into the soil. The nearest creek is several hundred feet away, and water soaking through the soil would take ten to a hundred years to traverse this distance. But the radioactive material would travel much more slowly; it would be effectively filtered out as the water passed through the soil, and would consequently take 100 times longer—a total of at least a thousand years—to reach the creek. We see that this alone gives a very high probability that the material will not get into the creek or lakes until its radioactivity is essentially gone.

But how likely is a release into the soil? The primary protection against this is the tank in which the waste is contained, but this will be discussed later. In addition, there are three further barriers keeping it from getting into the soil: First, the tanks are in a concrete vault which should contain the liquid. Second, the concrete vaults are surrounded by gravel, and there are pipes installed to pump water out of this gravel. If the radioactivity managed to get into this region it could still be pumped out through these pipes; there would be plenty of time—many weeks at least—to do this. Third, the entire cavity is in a highly impermeable clay that would take a very long time for the liquid to penetrate before reaching the ordinary soil. There is still one further barrier worthy of mention; the water flow in the creek is sufficiently small that during the ten or more years it would take the groundwater to reach it, a plant could be set up for removing the radioactivity from the creek water.

Some perspective on the danger of leakage into the ground may be gained from considering a Russian program in which more than twice the radioactive content of the West Valley storage tank was pumped down a well into the ground. This was done as an experiment to study movement of the radio-activity through the ground with a view to using this method for large scale high-level waste disposal.[37] At last report the results were consistent with expectations and their plans were proceeding.

Up to this point we have been assuming that the radioactive materials are in solution in the waste storage tank, but actually 95% of them are in a solid sludge which is lying on the bottom of these tanks. This sludge would be much less likely to get out through a leak, to penetrate the concrete vault, and to be transported through the ground with groundwater; even if it were dumped directly into Lake Erie, most of it would settle to the bottom, and even if it got into city water supplies, it would very probably be removed by the filtration system. The consequences of release into Lake Erie that we have given earlier are therefore probably ten times too high.

In summary, if there should be leakage from the tank, it would ordinarily be contained by the concrete vault. If it were not, it could be pumped out with the water which permeates the surrounding gravel. If this should fail, it would be contained for many years by the thick clay enclosing the entire cavity. When it eventually gets through to the surrounding soil, the movement of the radioactive materials is sufficiently slow that they would decay to virtually innocuous levels before reaching the creek. It would not be difficult to remove the radioactive material from the creek itself if this were necessary; if, as seems virtually certain, the material is delayed from reaching Lake Erie for at least 400 years, less than one fatality would be expected. If all else failed, any excess radioactivity in Lakes Erie and Ontario would be detected by routine monitoring operations, allowing precautions to be taken to protect public health.

How vulnerable is the tank to developing leaks? There is a very substantial system of protections against this, to be described in the next section. But it applies only to normal circumstances; what if there were a violent earthquake? A structural analysis indicates that even the most violent earthquake believed possible in that area would not rupture the waste storage tanks. (Such an earthquake is expected only once in 16,000 years.) One might consider sabotage of the tank with explosives; but the tank is covered with an eight-foot thickness of clay which would be extremely hazardous to dig through unless elaborate protective measures were taken. If the tank were successfully ruptured, all of the other protective barriers would remain intact, so in all likelihood no harm would result. Saboteurs have available many more inviting targets if their aim is to take human lives. As an example, they could easily kill thousands of people by introducing a poison gas into the ventilation system of a large building.

A very large bomb dropped from an airplane could reach the waste and vaporize it; if this happened, several hundred fatalities would be expected, but far more people would be killed if this bomb were dropped on a city. These same considerations apply to a possible strike by a large meteorite or the development of a volcano through the area. These latter events are, of course, extremely improbable.

Up to this point we have ignored the effects of the radioactive materials permeating the soil in the event of a leak. While in the soil, they could be picked up by plants and get into human food. How much of a hazard would this be? If all the radioactivity in the West Valley waste storage tank were now to become randomly distributed through the soil from the surface down to its present depth, if its behavior in soil is like that in average U.S. soil with the same percentage of land area used for farming, and if no protective action were taken, we would expect thirty fatalities.[35] If the situation were postponed for one hundred years, there would be three fatalities, and if it were postponed for more than 180 years, we would not expect any. Our assumption here that the material becomes randomly distributed through the soil up to the surface is probably a very pessimistic one. Also in the very unlikely event in which there could be a problem, it would easily be averted by checking food grown in the area for radiation and removing from the market any with excessive radioactivity.

In 1978, the U.S. Department of Energy set about deciding what to do about this waste tank. The simplest solution[35] would be to add cement mix to the tank to convert its contents into a large block of cement. This would eliminate any danger of leakage, making the principal danger that ground water could somehow penetrate successively through the clay barrier, the concrete vault, and the stainless steel tank wall to dissolve away some of this cement. Each of these steps would require a very long time period. For example, although the sides of swimming pools and dams are cement, we note that they aren't noticeably leached away in many years even by the soaking in water to which they are exposed; moreover, groundwater contact is more like a dampness than a soaking. If the material did become dissolved in ground water, all the barriers to getting into Lake Erie outlined above would still be in place and would have to be surmounted before any harm could be done. Even this remote danger could be removed by maintaining surveillance—periodically checking for water in the concrete vault and pumping it out if any should accumulate. The cost of converting to cement would be about $20 million, and a $15 million trust fund could easily provide all the surveillance one might desire for as long as anyone would want to maintain it.

If this were done, what would the expected health consequences be? I have tried to do risk analyses by assigning probabilities, and I find it difficult to obtain a credible estimate higher than 0.01 eventual deaths. It would be very easy to support numbers hundreds or thousands of times smaller.

However, this management option is not being taken. Instead the Department of Energy has decided to remove the waste from the tank, convert it to glass, and bury it deep underground in accordance with plans for future commercial high-level waste. This program will cost about $1 billion. Spending $1 billion to avert 0.01 deaths corresponds to $100 billion per life saved! This is going on at a time when the same government is turning down projects that would save a life for every $100,000 spent! That is our *real* waste problem.

One last item deserves mention here—the radiation exposure to workers in executing the plans described above. It turns out that this is larger in the billion dollar plan that was adopted than in the plan for conversion to cement, by an amount that would cause 0.02 deaths (i.e., a 2% chance of a single death) among the workers. Since this is more than 0.01 deaths to the public from the conversion to cement, the billion dollar plan is actually more dangerous!

I know the government officials who chose the billion dollar plan, and have discussed these questions with them. They are intelligent people trying to do their job well. But they don't view saving lives as the relevant question. Their job is to respond to public concern and political pressures. A few antinuclear zealots in the Buffalo area stirred up the public there with the cry "We want that dangerous waste out of our area"—why should any local people oppose them? Their congressional representatives took that message to Washington—what would they have to gain by doing otherwise? The Department of Energy officials responded to that pressure by asking for the billion dollar program—it wasn't hurting them; in fact, having a new billion dollar program to administer is a feather in their caps. Congress was told that a billion dollars was needed to discharge the government's responsibility in protecting the public from this dangerous waste—how could it fail to respond?

That is how a few antinuclear activists with little knowledge or understanding of the problem, induced the United States government to pour a billion dollars "down a rathole." I watched every step of the process as it went off as smooth as glass. And the antinuclear perpetrators of this mess are local heroes to boot.

LEAKING WASTE STORAGE TANKS

No discussion of nuclear waste problems could be complete without considering leaking waste storage tanks. We hear more stories about them than just about any other experience with waste handling.

When high-level waste is first isolated in a chemical reprocessing plant, it is stored in underground tanks for a few years before being converted into glass. One of these tanks might handle the waste accumulating from 50 power plants over several years, a very substantial quantity of radioactivity. This raises the question of the dangers from possible leaks in the tanks.[38]

This question arose very early in the history of nuclear energy. During World War II, the Hanford Laboratory was established in the desert of central Washington State to produce bomb material in reactors and separate it in chemical reprocessing plants, leaving the waste in underground storage tanks. These tanks were made of ordinary steel which was known to be readily corroded by the waste solution. It was therefore assumed that the tanks would eventually leak, releasing the radioactive waste into the soil. The thinking at that time was that this would be an acceptable situation—if the radioactive material was eventually to be buried underground anyway, why not let some of it get there through leaks in storage tanks? However, as public concern about radiation escalated over the years, this procedure became less acceptable, and in the 1960s the technology was changed. The new tanks were thereafter constructed of *stainless* steel which is much less easily corroded, and facilities were included to keep close track of any corrosion that might occur.* But more important, the new tanks were constructed with double walls. Thus if the inner wall develops a leak, the liquid will fill the space between the walls, thereby signaling the existence of the leak. In the meantime, the liquid is still contained by the outer tank wall, which leaves plenty of time to pump the contents into a spare tank.

These new tanks, which have been used in several locations for many years, including the West Valley tank described above, have had no problems, but some of the old single-wall tanks at Hanford have developed leaks, as was to be expected. On such occasions, the practice has been to pump the contents into a spare tank, but on one occasion in particular, a leak went undetected for seven weeks, resulting in discharge of a substantial quantity of radioactivity into the soil. Although there has been extensive adverse publicity from the incident (and there was irresponsible negligence involved) there have been no health consequences. Moreover, it is most difficult to imagine how there can ever be any, due to the following considerations.

All of the significant radioactive materials are now absorbed in the soil within a few feet of the tank, still 150 feet above the water table. As rain water occasionally percolates down through the soil, this radioactivity may

---

* Pieces of the material from which the tanks are constructed are kept suspended in the waste solution. Periodically one of these is withdrawn and studied for evidence of corrosion.

be expected eventually to reach the water table after several hundred years. Only then can it move horizontally, toward the Columbia River, about 10 miles away. However, groundwater flows very slowly in that region, requiring 800 years to cover that distance. The only radioactive material involved in the leak that will last more than a few hundred years is plutonium, which survives to some extent for a few hundred thousand years, but plutonium in ground water is constantly filtered out by the rock and hence travels 10,000 times more slowly than the water. It would therefore take nearly a million years to reach the river, by which time it would have decayed away. However, even if *all* of the plutonium involved in the leak (about 50 grams) were dumped into the river *now,* there is less than a 1% chance that even a single human health effect would result. While the analysis given here contains no controversial elements, and has never, to the best of my knowledge, been scientifically questioned, I have frequently heard word-of-mouth claims of detrimental health effects, both present and future, from the leaking tanks at Hanford. Rarely are such effects not at least hinted at in popular books and magazine articles about radioactive waste.

Some readers may be surprised by the statement above that 50 grams (1.8 ounces) of plutonium dumped in a river is so harmless. Part of the reason for this is that, as shown in the Appendix to Chapter 5, only one atom in 10,000, or 0.005 grams can be expected eventually to enter human stomachs. Another part of the reason is that when plutonium does enter a stomach, 99.99% of it is excreted within a few days, so it has little opportunity to irradiate the vital body organs.[39] The stories one hears about the high toxicity of plutonium (which will be discussed in Chapter 7) are all based on inhaling it into the lungs, rather than ingesting it with food or water.[40]

One final point should comfort anyone who may still be worried about future leaks in the Hanford waste storage tanks. Their contents are now being converted into a solid, which will greatly reduce corrosion rates, and will also largely eliminate any appreciable releases of radioactivity into the soil even if new openings should develop in the tank walls.

Waste Transport—When Radioactivity Encounters the Public[41,42]

Spent fuel must be shipped from reactors to reprocessing plants, and the high-level waste derived from it must be shipped to a repository for burial; therein lie substantial *potential* hazards. After all, inside plants the radioactivity is remote from the public, but in transport close proximity is unavoidable. Moreover, accidents are inherently frequent in transportation—half of all accidental deaths occur during the few percent of the time we spend traveling.

We will confine our attention to the shipping of spent fuel since that is where there has been the most experience; other high-level waste shipments will be handled analogously. One general safety measure is to delay shipping as long as practical to allow short-lived radioactivity to decay away: For spent fuel, the minimal delay has been six months. But the most important safety device is the cask in which the spent fuel is shipped. It typically costs a few million dollars, and one can well imagine that a great deal of protection can be bought for that kind of money.*

These casks have been crashed into solid walls at 80 miles per hour, and hit by railroad locomotives traveling at similar speeds, without any release of their contents. These and similar tests have been followed by engulfment in gasoline fires for 30 minutes and submersion in water for 8 hours, still without damage to the contents.[43] In actual practice, these casks have been used to carry spent fuel all over the country for more than 35 years. Railroad cars and trucks carrying them have been involved in all sorts of accidents, as might be expected. Drivers have been killed; casks have been hurled to the ground; but no radioactivity has ever been released, and no member of the public has been exposed to radiation as a consequence of such accidents.

If we try to dream up situations that would lead to serious public health consequences, we are limited by the fact that nearly all of the radioactivity is solid material, unable to leak out like a liquid or a gas. There is no simple mechanism for spreading it over a large area even if it did get out of the cask. In almost any conceivable situation, significant radiation exposure would be limited to people who linger for several minutes in the immediate vicinity of the accident; hence the number of people exposed would be relatively small.

When all relevant factors are included in an analysis, studies indicate[42] that with a very full nuclear energy program there would eventually be one death every *few thousand years* in the United States resulting from radioactivity releases in spent fuel transport accidents. Of course there would be many times that number of deaths from the normal consequences of these accidents.

There has been some discussion of the possibility that terrorists might blow up a spent fuel cask while it is being transported through a city. To

---

* One might wonder how it is feasible to spend millions of dollars on a shipping cask. The answer here again lies in the very small quantities of material involved. A large power plant might make 20 spent fuel shipments per year, accumulating a total of 100 days of cask use. Cask rental charges would therefore typically be much less than $1 million per year, which is less than one percent of the value of the plant's electrical energy production.

study this, Sandia National Laboratory carried out highly instrumented tests with high explosive devices. Their conclusion[44] was that even if several hundred pounds of high explosives were used on a cask traveling through downtown Manhattan, the total expected number of eventual deaths would be 0.2; that is, there is only a 20% chance that there would be a single death (from radiation induced cancer).

A more important effect is the slight radiation exposure to passers-by as trucks carrying radioactive waste travel down highways. Some gamma rays emitted by the radioactive material can penetrate the walls of the cask to reach surrounding areas (it is not practical to have as thick a shield wall on a truck as is used around reactors). Even with a full nuclear power program, however, exposures to individuals would be a tiny fraction of one millirem. The effects would add up to one death every few centuries in the United States.[42]

Some perspective on these results can be obtained by comparing them with impacts of other transport connected with energy generation. Gasoline truck accidents kill about 100 Americans each year,[45] and injure eight times that number. It has been estimated that coal-carrying trains kill about 1000 members of the public each year.[46] Clearly the hazards in shipment of radioactive waste are among the least of our energy-related transportation problems.

In spite of the very long and perfect record of spent fuel transport and its extremely small health effects, a great deal of public fear has been generated by media stories. Many municipalities, ranging from New York City down to tiny hamlets, have consequently passed laws prohibiting spent fuel transport through their boundaries. It does not take many such restrictions to cause extreme difficulty in laying out shipping routes. A New York City ordinance, for example, prevents any rail or truck shipments from Long Island to other parts of the United States.

The U.S. Department of Transportation, which has legal jurisdiction over interstate transport, has for several years been developing plans to rectify these problems. I hope that these plans reach fruition before too long, but in the meantime heavy financial losses are being incurred; as usual, the public is paying the bill.

A RADIOACTIVE WASTE ACCIDENT IN THE SOVIET UNION

While there is no evidence that anyone has ever been injured by radioactive waste in the United States, there has been wide publicity for an accident that may have involved such waste in the Soviet Union. The incident took place near the Kyshtym nuclear weapons complex in the Southern Ural Mountains during the Winter of 1957–1958. First reports[47] were from Z. A. Medvedev, a Soviet scientist now living in England, who pieced together information from rumors circulating among Russian scientists and from radiation contamination studies reported in the Russian literature. A second Soviet scientist,[48] now living in Israel, reported having driven through the area in 1960 and observing that the area was not inhabited; his driver told him that there had been a large explosion there. Medvedev theorized that this situation had resulted from a nuclear explosion of radioactive waste, and according to rumors he had heard, there had been many casualties and serious radioactive contamination over an area of several thousand square miles.

Following these reports, extensive studies were carried out at Oak Ridge National Laboratory,[49] and later at Los Alamos National Laboratory[50] using information culled from a thorough search of the Russian literature and U.S. government (CIA) sources. It was concluded that the incident resulted from some very careless handling of radioactive waste in the Russian haste to build a nuclear bomb, perhaps even storing high-level waste in open ponds. Several theories for the method of dispersal were considered by each group, but no definite conclusions could be drawn. The Oak Ridge study favored a theory of chemical explosion based on the assumption that the Russians used a process developed in this country but not used here because of its high explosive potential. The Los Alamos study favored a theory that the waste was spread by wind and ordinary water flow. Both groups concluded that it could not have been caused by a nuclear explosion of buried radioactive waste as had been conjectured by Medvedev, because the ratio of the quantities of various radioactive materials was grossly different from that produced by a nuclear explosion. Moreover, a large nuclear explosion requires that very special materials be brought together very rapidly, in something like one-millionth of a second, and that very little water be present,* whereas waste is normally dissolved in water, and the movement of materials is very slow. The U.S. researchers became quite convinced that the area contaminated was much smaller than Medvedev had estimated—rumor mills often exaggerate such figures and the definition of "contaminated" is subject to various interpretations. None of the accidents that seemed even remotely possible to the American researchers could have administered radiation doses approaching a lethal range. It is therefore highly dubious whether there were any casualties from radiation. Moreover, there is nothing in the Russian scientific literature on

---

* Water slows down neutrons and thereby reduces by a factor of 1000 the maximum rate of energy release.

medical effects of radiation to suggest that any had occurred in this incident. The chemical explosion theory favored by the Oak Ridge group, on the other hand, could have caused a number of fatalities; the chemical suspected of being used is ammonium nitrate, a shipload of which blew up in Texas City (Texas) in 1947, killing over 500 people.

The most definite conclusion of both U.S. studies was that no such accident could possibly occur with American waste-handling procedures—to answer that question was one of the prime reasons for these studies. None of the ingredients that could lead to either a nuclear or a chemical explosion is present in the U.S. technology.

The CBS program "60 Minutes" did a segment on the incident, featuring Medvedev and Ralph Nader. Medvedev implied that many thousands of people had been killed in the explosion, and Nader, who obviously had no information on the incident, spouted off about how the affair illustrates the great dangers of radioactive waste. The CBS crew, having spent two days in Oak Ridge, featured the work of that group in confirming that there had been such an accident, but it did not present their conclusions on its details. The Oak Ridge people repeatedly emphasized to the CBS crew that there is no way that such an incident could occur with U.S. procedures, but that point was not made in the TV presentation; on the contrary, Nader's statements strongly implied that such an accident could easily happen here.

It seems obvious to me that the question of whether there could be a similar accident with our waste management practices was paramount. All scientific analyses concluded that it could not, but the TV audience was given the impression by Nader that it could. The conclusion of the scientific analyses was kept from the public, and a politically inspired answer was presented instead. I can see no excuse for such blatant deception by the TV producers. Failure to transmit the scientific answer could not have been an oversight because that point was repeatedly emphasized to them by the Oak Ridge researchers. Surely they did not think the question "could it happen here?" was irrelevant. Several of my friends asked me that question over the next few weeks, and as soon as I walked into class the day after the program was aired, the first question my students asked was "Is such an incident possible with our waste management procedures?"

I am certain that failure to convey the answer of the scientific studies to this key question was not an oversight by the TV crew, but a purposeful attempt to conceal that information from the American public. Instead they gave the public Ralph Nader's implied answer that is scientifically indefensible. There was no other reason to have Nader on the show; he had no information on the incident, had never studied it, and has no scientific expertise on radioactive waste management. His only interest in the situation was to make politically inspired propaganda, and CBS-TV offered him this free opportunity to do so before its very large audience. (Nader's notorious unreliability on scientific aspects of nuclear power is reviewed in a recent book[51].) All the scientific community could do was grit its teeth in frustration.

SUMMARY

This chapter and the previous one presented detailed scientific analyses of the radioactive waste problem. The principal results of these analyses are contained in Table 1, with additional information in the section on waste transport. They show that the hazards of nuclear waste are thousands of times lower than those of wastes from other technologies. Nevertheless, the media have overpublicized and distorted the nuclear waste problem so grossly that billions of dollars are being wasted on it. Fortunately, even with all this waste of money, the cost of waste handling increases the price of nuclear electricity only by about one percent.

APPENDIX

We have stated that by far the most important radiation health effect of nuclear power is the lives saved by mining uranium out of the ground, thereby reducing the exposure of future generations to radon.[12] Let us trace through the process by which this is calculated.

The radon to which we are now exposed comes from the uranium and its decay products in the top one meter of U.S. soil, since anything percolating up from deeper regions has a good chance of decaying before reaching the surface. From the quantity of uranium in soil (2.7 parts per million[52]) and the land area of the U.S. (contiguous 48 states), it is straightforward to calculate that there are 66 million tons of uranium in the top meter of U.S. soil. As discussed previously, this is causing 10,000 deaths per year from radon now, and will continue to do so for something like 22,000 years, the time before it erodes away. This is a total of $(10{,}000 \times 22{,}000 =)$ 220 million deaths caused by 66 million tons of uranium, or 3.3 deaths per ton. As erosion continues, all uranium in the ground will eventually have its 22,000 years in the top meter of U.S. soil and will hence cause 3.3 deaths per ton.

In obtaining fuel for one nuclear power plant to operate for one year, 130 tons of uranium is mined out of the ground. This action may therefore be expected to avert ($180 \times 3.3 =$) 600 deaths. However, to be consistent we must also consider erosion of the mill tailings covers.[12] When this is taken into account, the net long-term effect of mining and milling is to *save* 500 lives.

On the short-term, 500-year perspective, we can ignore erosion, so uranium deep underground has no effect. However, about half of our uranium ore is surface mined, and about 1% of this ($0.005 \times 180 =$) 0.9 tons, is taken from the top one meter. This saves ($0.9 \times 3.6 =$) 3.3 lives over the next 22,000 years, or ($500/22,000 \times 3.3 =$) 0.07 lives over the next 500 years. We must subtract from this the lives lost by radon emission through the covered mill tailings, 0.003/year without the cover, $\times$ 1/200 to account for attenuation by the cover, $\times$ 500 years, $= 0.01$. Thus the net effect over the next 500 years of mining and milling uranium to fuel one power plant for one year is to *save* ($0.07 - 0.01 \cong$) 0.06 lives. That is the result used in our table.

We next consider the effects of radon from coal burning. Coal contains an average of 1.0 parts per million uranium[53]—some commercial coals contain up to 40 parts per million. When the coal is burned, by one way or another this is released into the top layers of the ground where it will eventually cause 3.3 deaths per ton with its radon emissions as shown above. The 3.3 million tons of coal burned each year by a large power plant release 3.3 tons of uranium, which is then expected to cause 3.3 deaths per ton, or a total of 11 deaths. These will be distributed over about 100,000 years, with some tendency for more of them to occur earlier; thus about 1% of them, or 0.11 deaths may be expected in the next 500 years. That is the figure in our table.

On a multimillion year time perspective,[14] the uranium in the coal would have eventually reached the surface by erosion even if the coal had not been mined; these 11 deaths would therefore have occurred eventually anyhow, and hence can be discounted. However, the *carbon* in the coal, its main constituent, is burned away, whereas if the coal had been left in the ground and eventually reached the surface by erosion, this carbon would have prevented radon emission since it contains no uranium (the uranium in the coal has already been accounted for). Since that carbon is missing, its time near the surface will be taken by other rock which does contain uranium, 2.7 parts per million on an average, or ($3.3 \times 2.7 =$) 9 tons in the 3.3 million tons of rock which replaces the carbon. Multiplying this by 3.3 deaths per ton of uranium gives 30 deaths, the result we have used.

REFERENCE NOTES

1. A convenient review of radon properties and behavior is given in B. L. Cohen, "Radon: Characteristics, Natural Occurrence, Technological Enhancement, and Health Effects," *Progress in Nuclear Energy,* **4,** 1 (1979).
2. F. E. Lundin, J. K. Wagoner, and V. E. Archer, "Radon Daughter Exposure and Respiratory Cancer: Quantitative and Temporal Aspects," U.S. Department of HEW-NIOSH-NIEHS Joint Monograph No. 1 (1971).
3. U.S. National Academy of Sciences Committee on Biological Effects of Ionizing radiation, "The Effects on Populations of Exposure to Low Levels of Ionizing Radiation" (1980).
4. B. L. Cohen, "Radon Daughter Exposure to Uranium Miners," *Health Physics,* **42,** 449 (1982).
5. The United Nations Scientific Committee on Effects of Atomic Radiation, "Sources and Effects of Ionizing Radiation," United Nations, New York (1977).
6. A. J. Breslin and A. C. George, "Radon Sources, Distribution, and Exposures in Residential Buildings," *Transactions of the American Nuclear Society,* **33,** 150 (1979); J. Rundo, F. Markun, and N. J. Plondke, "Observations of High Concentrations of Radon in Certain Houses," *Health Physics,* **36,** 729 (1979); G. A. Franz, Colorado State Department of Health, private communication (1979); Health and Rehabilitative Services Department, State of Florida, "Study of Radon Daughter Concentrations in Structures in Polk and Hillsborough Counties" (1978); E. Stranden, L. Berteig, and F. Ugletveit, "A Study of Radon in Dwellings," *Health Physics,* **36,** 413 (1979).
7. B. L. Cohen, "Health Effects of Radon from Insulation of Buildings," *Health Physics,* **39,** 937 (1980).
8. C. D. Hollowell, M. L. Boegel, J. G. Ingersoll, and W. W. Nazaroff, "Radon-222 in Energy Efficient Buildings," *Transactions of the American Nuclear Society,* **33,** 148 (1979).
9. U.S. Environmental Protection Agency, "Environmental Analysis of the Uranium Fuel Cycle," Report EPA-520/9-73-003B (1973); R. O. Pohl, "Health Effects of Radon-222 from Uranium Milling," *Search,* **7,** 345 (1976); C. C. Travis, *et al.,* "Radiological Assessment of Radon-222 Released from Uranium Mills and Other Natural and Technologically Enhanced Sources," U.S. Nuclear Regulatory Commission Report NUREG/CR-0573; U.S. Nuclear Regulatory Commission, Final Generic Environmental Impact Statement on Uranium Milling, NUREG-0706.
10. B. L. Cohen, "Health Effects of Radon Emissions from Uranium Mill Tailings," *Health Physics,* **42,** 695 (1982).
11. U.S. Code of Federal Regulations, Title 10, Part 40, Appendix A.
12. B. L. Cohen, "The Role of Radon in Comparisons of Effects of Radioactivity Releases from Nuclear Power, Coal Burning, and Phosphate Mining," *Health Physics,* **40,** 19 (1981).

13. E. D. Goldberg *et al.*, "Marine Chemistry," in *Radioactivity in the Marine Environment*, National Academy of Sciences (1971).
14. B. L. Cohen, "Health Effects of Radon from Coal Burning," *Health Physics*, **42**, 725 (1982).
15. P. D. Moskowitz *et al.*, "Photovoltaic Energy Technologies: Health and Environmental Effects Document," Brookhaven National Laboratory Report BNL-51284 (1980); P. Masser, "Low Cost Structure for Photovoltaic Arrays," Sandia Laboratory Report, SAND-79-7006 (1979); B. L. Cohen, "Consequences of a Linear, No-Threshold Dose–Response Relationship for Chemical Carcinogens," *Risk Analysis*, **1**, 267 (1981).
16. J. W. Gofman and A. R. Tamplin, *Poisoned Power* (Rodale Press, Emmaus, Pa. 1971).
17. Code of Federal Regulations, Title 10, Part 50, Appendix I. U.S. Nuclear Regulatory Commission, Regulatory Guides 1.109-1.113 (1976). These specify expenditure of $1000/man-rem; dividing by the fatal cancer risk from Chapter 2, 1/8000 per man-rem gives $8 million per fatal cancer averted.
18. Code of Federal Regulations, Title 40, Part 190.
19. Kentucky Department of Human Resources, "Six Month Study of Radiation Concentrations and Transport Mechanisms at the Maxey Flats Area of Fleming County, Kentucky" (1974); G. L. Meyer, Preliminary Data on the Occurrence of Transuranium Nuclides in the Environment at the Radioactive Waste Burial Site, Maxey Flats, Kentucky;" EPA Report, Office of Radiation Programs (1976).
20. J. Hardin, Kentucky Department of Human Resources, private communication.
21. J. Razor (Maxey Flats Environmental Protection Officer), private communication.
22. U.S. House of Representatives Subcommittee on Conservation, Energy, and Natural Resources, William B. Moorhead, Chairman, Hearings on February 23 (1976).
23. J. M. Matuszek, F. V. Strnisa, and C. F. Baxter, "Radionuclide Dynamics and Health Implications for the NY Nuclear Service Center's Radioactive Waste Burial Site," International Atomic Energy Agency paper IAEA-SM-207/59 (1976).
24. National News Council, *Cohen et al. against NBC News*, filed February 23 (1977).
25. W. F. Naedle, "Nuclear Grave is Haunting KY," *Philadelphia Evening Bulletin*, May 7 (1979).
26. U.S. Code of Federal Regulations, Title 10, Part 61.
27. *Federal Register*, "Licensing Requirements for Land Disposal of Radioactive Waste," Vol. **46** (142), 38096ff July 24 (1981).
28. Nuclear Regulatory Commission, Draft Environmental Statement on 10CFR61, NRC Document NUREG-0782 (1981).
29. B. L. Cohen, "A Generic Probabilistic Risk Assessment for Low Level Waste Burial Grounds," (submitted for publication, *Nuclear and Chemical Waste Management*).
30. H. J. Rosler and H. Lange, *Geochemical Tables* (Elsevier Publ. Co., Amsterdam, 1972).

31. International Commission on Radiological Protection , *Report of the Task Group on Reference Man,* ICRP Publication No. 23 (Pergamon Press, New York, 1975).
32. Nuclear Regulatory Commission, Data Base for Radioactive Waste Management, NRC Document NUREG/CR-1759 (1981).
33. U.S. Energy Research and Development Administration (ERDA), "Alternatives for Long Term Management of Defense High-Level Radioactive Waste," Document ERDA-77-42/1 (1977).
34. L. R. Brown, "World Food Resources and Population," Population Reference Bureau (1981).
35. Western New York Nuclear Service Center Study, U.S. Department of Energy Report TID-28905-2 (1980).
36. B. L. Cohen, "The Situation at West Valley," *Public Utilities Fortnightly,* Sept. 27, (1979), p.26.
37. V. I. Spitsyn *et al.,* International Atomic Energy Agency paper IAEA-CN-36/345 (1977).
38. U.S. Atomic Energy Commission, "Proposed Final Environmental Statement, Liquid Metal Fast Breeder Reactor Program," Document WASH-1535, December (1974).
39. International Commission on Radiological Protection (ICRP), *Limits for Intakes of Radionuclides by Workers,* ICRP Publication No. 30 (Pergamon Press, New York, 1979).
40. W. J. Bair and R. C. Thompson, "Plutonium: Biomedical Research," *Science,* **183,** 715 (1974).
41. U.S. Department of Energy, "Environment Impact Statement, Management of Commercially Generated Radioactive Waste," Document DOE/EIS-0046F, October (1980).
42. Nuclear Regulatory Commission, "Final Environmental Statement on the Transportation of Radioactive Material by Air and Other Modes," Document NUREG-0170 (1977).
43. R. M. Jefferson and H. R. Yoshimura, "Crash Testing of Nuclear Fuel Shipping Containers," Sandia National Lab Report SAND 77-1462 (1978).
44. R. P. Sandoval and G. J. Newton, "A Safety Assessment of Spent Fuel Transportation Through Urban Regions," Sandia National Laboratory Report SAND 81-2147 (1981); R. P. Sandoval, "Safety Assessment of Spent Fuel Transport in an Extreme Environment," *Nuclear and Chemical Waste Management,* **3,** 5 (1982).
45. A. Parachini, "Study Finds Tanker Trucks a Highway Danger," *Los Angeles Times,* May 21 (1982).
46. C. Bliss *et al.,* "Accidents and Unscheduled Events Associated with Non-nuclear Energy Resources and Technology," Mitre Corp-EPA Report PB-265 398, p. 65 (1981) P. Gleich, "Health and Safety Effects of Coal Transportation," *Energy,* 614 (1981).
47. Z. A. Medvedev, *New Scientist,* **72,** 264 (1976); **72,** 692 (1976); **74,** 761 (1977); **76,** 352 (1977).
48. L. Tumerman, quoted by W. E. Farrell, *New York Times,* December 9, 8 (1976).

49. J. R. Trabalka, L. D. Eyman, and S. I. Auerbach, "Analysis of the 1957–58 Soviet Nuclear Accident," *Science,* **209,** 345 (18 July 1980).
50. D. M. Soran and D. B. Stillman, "An Analysis of the Alleged Kyshtym Disaster," Los Alamos National Lab Report LA-9217-MS (1981).
51. S. McCracken, *The War Against the Atom* (Basic Books, New York, 1982).
52. *Encyclopedia of Science* (article on Uraniuim) (McGraw-Hill, New York, 1971).
53. J. Cavallaro, A. W. Deurbrouck, H. Schultz, and G. A. Gibbon, "Washability and Analytical Evaluation of Potential Pollution from Trace Elements in Coal," Environmental Protection Agency Report EPA-600/7-78-038, March (1978); H. J. Gluskoter, "Trace Elements in Coal; Occurrence and Distribution," EPA Report EPA-600/7-77-064 (1977); C. T. Ford, "Evaluation of the Effect of Coal Cleaning on Fugitive Elements," Bituminous Coal Research Lab Draft Report L-1082 (1980).

Chapter 7 / PLUTONIUM AND BOMBS

The very existence of plutonium is often viewed as the work of the devil.* As the most important ingredient in nuclear bombs, it may someday be responsible for killing untold millions of people, although there are substitutes for it in that role if it did not exist. If it gets into the human body, it is highly toxic. On the other hand, its existence is the only guarantee we have that this world can obtain all the energy it will ever need *forever* at a reasonable price. In fact I am personally convinced that citizens of the distant future will look upon it as one of God's greatest gifts to humanity. Between these extremes of good and evil is the fact that if our nuclear power program continues to be run as it is today, the existence of plutonium will have no relevance to it except as a factor in technical calculations.

Clearly there are several different stories to tell about plutonium. We will start with the future benefits, then discuss the weapons connection, and conclude with the toxicity question.

---

* It is sometimes said that it was named with that in mind, but it was actually named for the planet Pluto.

FUEL OF THE FUTURE

As uranium occurs in nature, there are two types, U-235 and U-238, and only the former, which is less than 1% of the mixture, can be burned (i.e., undergo fission) to produce energy. Thus present-day power reactors burn less than 1% of the uranium that is mined to produce their fuel. This sounds wasteful but it makes sense economically, because the cost of the raw uranium at its current price represents only 5% of the cost of nuclear electricity (cf. Appendix). However, there is only a limited amount of ore from which uranium can be produced at anywhere near the current price, perhaps enough to provide lifetime supplies of fuel needed by all nuclear power plants built up to the year 2005. Beyond that, uranium prices would escalate rapidly, doubling the cost of nuclear electricity within several decades.

Fortunately, there is a solution to this problem. The fuel for present-day American power plants is a mixture of U-238 and U-235. As the reactor operates, some of the U-238, which cannot burn, is converted into plutonium. This plutonium can undergo fission and thus serve as a nuclear fuel. In fact some of it is burned while the fuel is in the reactor, enough to account for one-third of the reactor's total energy production. But some of it remains in the spent fuel from which it can be extracted by chemical reprocessing. This plutonium could be burned in our present power reactors, but an alternative is to use it in another type of reactor, the *breeder*, whose fuel is a mixture of plutonium and uranium (U-238). Much more of the U-238 in the breeder is converted to plutonium than in our present reactors, more than enough to replace all of the plutonium that is burned. Thus a breeder reactor not only generates electricity, but it produces its own plutonium fuel with extra to spare. It only consumes U-238, which is the 99 + % of natural uranium that cannot be burned directly; it therefore provides a method for indirectly burning this U-238. With it, nearly all of the uranium, not less than 1% as in present type reactors, is eventually burned to produce energy. About a hundred times as much energy is thus derived from the same initial quantity. That means that instead of lasting only for about 50 years, our uranium supply will last for thousands of years. As a bonus, the environmental and health problems from uranium mining and mill tailings will be reduced a hundredfold. In fact, all uranium mining could be stopped for about 200 years while we use up the supply of U-238 that has already been mined and is now in storage.

Deriving 100 times as much energy from the same amount of uranium fuel means that the raw fuel cost per kilowatt-hour of electricity produced is reduced correspondingly. In fact the fuel costs per unit of useful energy generated in a breeder reactor are equivalent to those of buying gasoline at a price of 40 gallons for a penny! (cf. Appendix). Instead of contributing 5% to the price of electricity as in present type reactors, the uranium cost then contributes only 0.05% in a breeder reactor. If supplies should run short, we can therefore afford to use uranium that is twenty times more expensive, for even that would raise the cost of electricity by only (20 × .05 =) 1%. How much uranium is available at that price?

The answer is effectively *infinite* because it includes uranium separated out of sea water.[1] The world's oceans contain five billion tons of uranium, enough to supply all the world's electricity through breeder reactors for several million years. But in addition, rivers are constantly dissolving uranium out of rock and carrying in into the oceans, renewing the oceans' supply at a rate sufficient to provide 25 times the world's present total electricity usage.[2] In fact, breeder reactors operating on uranium extracted from the oceans could produce all the energy mankind will ever need* without the cost of electricity increasing by even one percent due to raw fuel costs.

The fact that raw fuel costs are so low does not mean that electricity from breeder reactors is very cheap. The technology is rather sophisticated and complex, involving extensive handling of a molten metal (liquid sodium) that reacts violently if it comes in contact with water or air. Largely as a result of the safety precautions required by this problem, the cost of electricity from the breeder will be something like 30% higher at today's uranium prices than that from reactors now in use.[3] Nevertheless, France, England, and the Soviet Union are in advanced stages of developing breeder reactors, and several other countries, including Germany and Japan, are involved to a lesser degree. The American program was at the forefront 15 years ago, but it now lags far behind as political fights are waged annually over whether the program should be scrapped.

On the surface, the opposition to breeder reactors is based on the fact that the United States has plentiful uranium resources, which is interpreted to mean that there is no need for the breeder at this time. That is a questionable argument for several reasons. First, even if development goes forward at the hoped-for pace, it will be many years before the first commercial breeder can become operational and many more before its use would become widespread; it is better to start up any new technology slowly, allowing the "bugs" to be worked out before a large number of plants is built. Second, we are not that certain about our uranium resources; they may be substantially below current estimates. Having the breeder reactor ready would be a cheap insurance policy

---

* The earth is expected to last for about 5 billion years before it becomes a molten mass due to changes in the sun. The uranium supply is adequate for that time period.[2]

against that eventuality, or against any sharp increase in uranium prices for whatever the reason. And third, our breeder reactor development program has substantial momentum with lots of scientists, engineers, and technicians deeply involved. It would be much more efficient to carry the program to completion now than to stop it, allow these people to become scattered, and then start over with a new team of personnel later.

Not far beneath the surface, there is substantial opposition to the breeder because of distaste for plutonium and general opposition to nuclear power. There are also some fears about the safety of breeder reactors, but experts on that subject (of which I am not one) maintain that they are extremely safe, and even safer than present reactors.[3-8] They have the important safety advantage of operating at normal pressure rather than at very high pressure as is the case for present reactors. There are therefore no forces tending to enlarge cracks or to blow the coolant out of the reactor (this is the "blowdown" discussed in Chapter 3).

Breeder reactor development represents the principal current U.S. government investment in nuclear power. It is often claimed that our government is heavily subsidizing commercial nuclear power, but this is not true; it is contrary to explicitly declared policy. Some aspects of our nuclear power program are by law conducted by the government, notably disposal of high-level waste and isotope separation of uranium, i.e., increasing the ratio of U-235 to U-238 in fuel material. But the full costs for these services, as well as for government regulation of the industry and other services, are charged to the utilities and through them, to the cost of electricity. The cost of waste disposal is paid by a 0.1 cent/kilowatt-hour tax on nuclear electricity, which is considerably more than it is now planned to spend. The isotope separation is carried out in plants constructed in the 1950s at very low cost by present standards to produce materials for military applications, but the charges for their services to utilities are computed as though they were constructed today. Government policy is to finance research and development of future nuclear technology, but operation of the industry with present technology is in no sense subsidized. On the contrary, it is very heavily penalized by the gross overconcern for safety and environmental impacts and the regulatory turbulence this engenders (cf. Chapter 8). This situation contrasts sharply with the case of solar energy where the U.S. government subsidizes 40% of the cost to all users through a tax credit, and where some states provide up to 30% additional, leaving only 30% for the user to pay.

A key part of the breeder reactor cycle is the reprocessing of spent fuel to retrieve the plutonium. In fact, this must be done with the spent fuel from present reactors in order to obtain the plutonium necessary to fuel the first generation of breeder reactors. As long as there is no reprocessing, the plutonium occurs only in spent fuel, where it is so highly dilute (1/2 of 1% of the total) that it is unusable for any of the purposes usually discussed. Moreover, spent fuel is so highly radioactive (independently of its plutonium content) that it can only be handled by large and expensive remotely controlled equipment. It therefore cannot be readily stolen or used under clandestine conditions. Without reprocessing, there is no use for plutonium, for good or evil.

It should also be recognized that plutonium plays only a minor role in waste disposal problems, and a negligible role in reactor accident scenarios. Thus, as long as there is no reprocessing, which is the present status in the United States commercial nuclear power program, plutonium issues have no direct relevance to the acceptability of nuclear power.

However, it is my personal viewpoint that it is *immoral* to use nuclear power without reprocessing spent fuel. If we were simply to irretrievably bury it, we would consume all the rich uranium ores within about 50 years. This would deny future citizens the opportunity of setting up the breeder cycle, the only reasonably low cost source of energy for mankind's future of which we can be certain. By such action, our generation might well go down in history as the one that denied mankind the benefits of cheap energy for millions of years, a fitting reason to be eternally cursed. On the other hand, if we develop the breeder reactor, we may go down in history as the generation that solved the world's energy problems for all time. Mankind might well remember and bless us for millions of years.

Unfortunately, the people in control are not worried about the long-range future of mankind. People in the nuclear power industry are concerned principally about the next 30 or 40 years, and politicians rarely extend their considerations even that far into the future. Whether or not we do reprocessing will have little impact over these time periods; thus the prospects for early reprocessing are questionable.

The situation was very different only a few short years ago. A large reprocessing plant capable of servicing most of the power plants now operating in the United States was constructed near Barnwell, South Carolina, by a consortium of chemical companies. The main part of the plant, costing $250 million, was completed in 1976, but two add-ons that would have cost about $130 million were delayed by government indecision. Since the add-ons would not be needed for several years, it was expected that the main part of the plant could be put into immediate operation.

At that critical point, the U.S. Government decreed an indefinite deferral of commercial reprocessing. The reason for the decree involved our national policy on discouraging proliferation of nuclear weapons, which will be discussed later in this chapter, but from the viewpoint of the plant owners, it was a disaster. They had been strongly encouraged to build the plant by government agencies—for example, Federally owned land was made available to them for purchase—and every stage of the planning was done in close consultation with those agencies. They had scrupulously fulfilled their end of the bargain, laying out a large sum of money, and now they were left with a plant earning no income.

By the time the Reagan Administration withdrew the decree forbidding reprocessing five years later, the owners had lost heart in the project, and were unwilling to provide the money, now increased to over $200 million, to provide the add-ons. A new consortium has indicated that they might be willing to take over the project if the Government would (1) provide assurances that any additional money invested would be compensated if the project were again terminated by political decree, and (2) guarantee to purchase the plutonium it produces. The latter requirement is necessary because the Barnwell plant was originally built with the understanding that utilities could purchase the plutonium to fuel present reactors, but the government has not taken action to allow this and probably will never do so. It is now widely agreed that it would be better to save the plutonium for breeder reactors. Since there are no commercial breeder reactors in the United States and will probably not be any for many years, this leaves the Government as the only customer for the plutonium from the reprocessing plant. Whether it provides the requested guarantees, and whether this will then lead to the required business arrangements to allow the Barnwell plant to operate, remains to be seen.

Aside from the idealistic considerations of providing energy for future generations, the main driving force behind getting reprocessing plants into operation is its contribution to waste management. Power plants are having difficulty in storing all of the spent fuel they are discharging; reprocessing gives them an outlet for it. Furthermore, the amount of material to be buried is very much reduced if the uranium is removed in reprocessing. There is also considerably more security in burying high-level waste converted to glass and sealed inside a corrosion-resistant casing, than in burying unreprocessed spent fuel which would have to be encased in asphalt or some similar material. In addition, there is an economic advantage since the value of the plutonium and uranium separated out (not to mention the money saved in waste management) exceeds the cost of reprocessing.

On the other hand, there has been, and continues to be, strong opposition to reprocessing. There have been well-publicized attacks on its environmental acceptability, ignoring the contrary evidence in the scientific literature in favor of "analyses" by "environmental groups" tailored to reach the desired conclusion. There have been widely publicized economic analyses of unspecified origin claiming that reprocessing is a money-losing proposition, although the real professionals in the business continue to find that it is economically advantageous.[9] There was a considerable amount of publicity for a paper issued by the Department of Energy claiming that the Barnwell plant is technically flawed,[10] but it turned out that it was by a nonscientist with little experience in the field who had never visited the plant and was confused over differences between reprocessing fuel from present power reactors and breeder reactors; the paper had accidently slipped through the Department of Energy reviewing process and was disavowed and strongly critiqued by the head of the Division that had issued it.[11]

A major part of this opposition to reprocessing comes from those opposed to nuclear power in general for political or philosophical reasons. They probably realize that it is too late to stop the present generation of reactors, but if they can stop reprocessing, nuclear power can have no long-term future. However, the most important opposition to reprocessing stems from its possible connection to nuclear weapons. If there is a connection between nuclear electricity and nuclear explosives, reprocessing is the bottleneck through which it must pass. We now turn to a discussion of that matter.

PROLIFERATION OF NUCLEAR WEAPONS

Everyone agrees that nuclear weapons can have very, very horrible effects and that it is exceedingly important to avert their use against human targets. One positive step in this direction is to minimize the number of nations that have them available for use; that is, to avoid the proliferation of nuclear weapons. To what extent do nuclear power programs frustrate that goal?

If a nation has a nuclear power reactor and a reprocessing plant, it could reprocess the spent fuel from the reactor to obtain plutonium, and then use that plutonium to make bombs. On the other hand, there are much better ways for nations to obtain nuclear weapons. There are two* practical fuels for nuclear fission bombs: uranium-235 (U-235), which occurs in nature as less than 1% of normal uranium from which it must be removed by a process known as "isotope separation," and plutonium, which can be produced in nuclear reactors and converted into usable form through reprocessing. Both

---

* There is a third possible fuel, uranium-233, but considering it would unduly complicate the discussion here.

the isotope separation and the reactor-reprocessing methods are used by all five nations known to have nuclear weapons, namely, the United States, the Soviet Union, Britain, France, and China.* Either method can produce effective bombs, although the best bombs use a combination of both.

However, there is a subtle aspect to producing plutonium by the reactor-reprocessing method and to explain it we must divert our discussion briefly to describe how a plutonium bomb works. There are two stages in its operation: first, there is an *im*plosion in which the plutonium is blown together and powerfully compressed by chemical explosives which surround it, and then there is the *ex*plosion in which neutrons are introduced to start a rapidly escalating chain reaction of fission processes which release an enormous amount of energy very rapidly to blow the system apart. All of this takes place within a millionth of a second, and the timing must be precise—if the explosion phase starts much before the implosion process is completed, the power of the bomb is greatly reduced. In fact, one of the principal methods that has been considered for defending against nuclear bombs is to shower them with neutrons to start the explosion early in the implosion process, thereby causing the bomb to fizzle. For a bomb to work properly, it is important that no neutrons come upon the scene until the implosion process approaches completion.

Plutonium fuel, plutonium-239, is produced in a reactor from uranium-238, but if it remains in the reactor it may be converted into plutonium-240 which happens to be a prolific emitter of neutrons. In a power plant, the fuel typically remains in the reactor for three years, as a consequence of which something like 30% of the plutonium produced comes out as plutonium-240. If this material is used in a bomb, the plutonium-240 produces a steady shower of two million neutrons per second,[12] which on an *average* would reduce the power of the explosion tenfold, but might cause a much worse fizzle. In short, a bomb made of this material, known as "reactor grade plutonium," has a relatively low explosive power and is highly unreliable. It is also more difficult to design and construct.

A much better bomb fuel is "weapons grade plutonium," produced by leaving the material in a reactor for only about 30 days. This reduces the amount of plutonium-240 and hence the number of neutrons showering the bomb tenfold.

One might consider trying to use a power reactor to produce weapons grade plutonium by removing the fuel for reprocessing every 30 days, but

* India has also exploded a nuclear device but claims that it was for nonmilitary purposes.

this would be highly impractical because fuel removal requires about a 30-day shutdown. Moreover, the fuel for a power reactor is very expensive to fabricate because it must operate in a very compact geometry at high temperature and pressure to produce the high-temperature, high-pressure steam needed to generate electricity.

It is much more practical to build a separate plutonium *production reactor* designed not to generate electricity but rather to provide easy and rapid fuel removal in a spread out geometry with fuel that is cheap to fabricate because it operates at low temperature and normal pressure. Moreover it can use natural uranium rather than the very expensive enriched uranium needed in power reactors. For a given quantity of fissile material, the former contains four times as much of the uranium-238 from which plutonium is made, hence producing four times as much plutonium. A plutonium production reactor costs less than one-tenth as much as a nuclear power plant,[13] and could be designed and built much more rapidly. All of the plutonium for all existing military bombs has been produced in this type of reactor.

Another alternative would be to use a *research reactor,* designed to provide radiation for research applications* rather than to generate electricity. At least 45 nations now have research reactors, and in at least 25 of these there is a capability of producing enough plutonium to make one or more bombs every two years. Research reactors are usually designed with lots of flexibility and space, so it would not be difficult to use them for plutonium production.

A plant for generating nuclear electricity is by necessity large and highly complex with most of the size and complexity due to reactor operation at very high temperature and pressure, the production and handling of steam, and the equipment for generation and distribution of electricity. It would be impossible to keep construction or operation of such a plant secret. Moreover, only a very few of the most technologically advanced nations are capable of constructing one. No nation with this capability would provide one for a foreign country without requiring elaborate international inspection to assure

---

* Perhaps the most important application of research reactors is producing radioactive isotopes for use in medicine, agriculture, industry, education, etc. Another is to provide neutron beams for studying properties of materials ranging from metals to biological molecules. They provide the most sensitive means available for measuring tiny quantities of material such as the gunpowder residue on a criminal's hand from firing a gun, or the mercury in fish. There are many other applications of research reactors ranging from basic physics and chemistry to practical engineering and biology.

that its plutonium is not misused. A production or research reactor, on the other hand, can be small and unobtrusive. It has no high pressure or temperature, no steam, and no electricity generation or distribution equipment. Almost any nation has, or could easily acquire, the capability of constructing one, and it probably could carry out the entire project in secret. There would be no compulsion to submit to outside inspection.

In view of the above considerations, it would be completely illogical for a nation bent on making nuclear weapons to obtain a power reactor for that purpose. It would be much cheaper, faster, and easier to obtain a plutonium production reactor; the plutonium it produces would make much more powerful and reliable bombs with much less effort and expense.

The only reasonable scenario in which power reactors might be used involves a nation that decided it needed nuclear weapons *in a hurry*. In such a situation, one or two years could be saved if a power reactor were available and a production or large research reactor were not.[13] However, nearly all nations that have a power reactor also have research reactors. Moreover, it would be most unusual for this time saving to be worth the sacrifice in weapons quality.

But obtaining plutonium is not the only way to get nuclear weapons. The other principal method is to develop *isotope separation* capability. Nine nations now have facilities for isotope separation,[13] and others would have little difficulty in acquiring it. A plant for this purpose, costing $20–$200 million could provide the fuel for 2–20 bombs per year, and could be constructed and put into operation in 3–5 years.[13] The product material would be very easy to convert into excellent bombs, much easier than making a plutonium bomb even with weapons grade plutonium.

This assessment is based on present technology, but several new, simpler, and cheaper technologies for isotope separation are under development and will soon be available. They will make the isotope separation route to nuclear weapons even more attractive. There are also new technologies under consideration for producing plutonium without reactors which may make that route more attractive.

The way I like to explain the problem of nuclear weapons proliferation is to consider three roads to that destination: (1) isotope separation, (2) plutonium production with research or production reactors, and (3) plutonium production in power plants, with (2) and (3) requiring reprocessing. The first two roads are much more attractive than the third from various standpoints; they are like super highways, while the third is like a twisting back country road. In this analogy, how important is it to block off the third road while leaving the first two wide open? The link between nuclear power and proliferation of nuclear weapons is a weak and largely insignificant one.*

But that is certainly not the impression given by the media. The great majority of stories about nuclear weapons proliferation involves nuclear power plants. They generally give the impression that without nuclear power there would be no proliferation problem. They rarely differentiate between a power reactor and other types more suitable for making bombs. I believe most Americans think that the Iraqi reactor destroyed by an Israeli air raid was a nuclear power plant, when in fact it was a large research reactor.

Even though nuclear power plants are only a minor source of weapons proliferation, nobody is saying that elaborate precautions should not be taken to see that the plutonium in them is not used for that purpose. The programs for dealing with that problem are known as *safeguards*. They are administered by the International Atomic Energy Agency (IAEA) based in Vienna. IAEA has teams of inspectors trained and equipped to detect diversion of plutonium. They have ready access to all nuclear power plants, reprocessing plants, and other facilities involved in handling plutonium† in nations that adhere to that treaty. There has been an impressive development in techniques and equipment for carrying out their mission over the last few years. For example, the Barnwell reprocessing plant has developed an automatic computer-controlled system that will give a warning within less than an hour if any plutonium in that plant should not be where it is supposed to be. With such measures and IAEA inspectors on the scene or making unannounced visits, it would be very difficult for a nation to divert plutonium from its nuclear power program without the rest of the world knowing about it long before the material can be converted into bombs.

These safeguards would be much easier to circumvent with a production or research reactor or with an isotope separation plant. These are much smaller operations with far less support needed from foreign suppliers; it would not be difficult to build them clandestinely. The IAEA safeguards system thus does much more to block off the twisting back country road than the super highways.

---

* There would be a somewhat stronger link—but still not a very strong one—with breeder reactors. However, it would not be difficult to limit breeder reactors to nations in which there is not an issue of weapons proliferation. In any case, no such nation will have a breeder reactor for the next several decades.
† The principal other facility would be for fabricating plutonium fuel for use in breeder reactors or perhaps in present reactors if the price of uranium should increase sharply.

NONPROLIFERATION POLITICS

One would have thought that that would be enough attention paid to the back country road, but the Carter Administration saw fit to go a step further. It decided to try to prevent the acquisition of reprocessing technology by non -nuclear-weapons nations. As you may recall, reprocessing is a bottleneck that must be passed if nuclear power plants are to be used to make bomb materials; thus the goal of the government was, in principle, a desirable one. However, the method for implementing it was disastrous.

At that time (1977), Germany was completing a deal to set up a repro- cessing plant in Brazil, Japan was building a plant, and France was negotiating the sale of plants to Pakistan and Korea. The Carter goal was to stop these activities through moral and political pressure. To set the moral tone for this effort—essentially to "show that our heart is in the right place"—he decided to defer indefinitely the reprocessing of commercial nuclear fuel in the United States.* This was the move that prevented the Barnwell plant from operating.

There were several problems with this approach. One was that the U.S. government continued to do reprocessing in its military applications program, which was something of a dilution of the high moral tone being advertised. Another was that Germany had just won the Brazilian contract after stiff bidding competition with U.S. firms. The Germans therefore interpreted the U.S. initiative as "sour grapes" over the loss of business. But a much bigger problem arose from the political pressure used: the United States delayed and threatened to stop shipments of nuclear fuel to nations that would not co- operate.

American manufacturers had built up a thriving export business of selling reactors to countries all over the world. Part of the deals was a guaranteed future supply of fuel for the reactors; this involved U.S. government partic- ipation in the contracts because it possessed the only large-scale facilities for isotopic enrichment of uranium. These sales contracts had no clauses allowing interruption of the fuel supply—no one would spend hundreds of millions of dollars for a power plant without a guaranteed fuel supply—so the delays and threatened withholding of shipments by the United States represented a direct and illegal breach of contract. Even nations with no interest in reprocessing were deeply upset by the very principle of this action. I remember sitting in

---

* Actually, Mr. Carter raised the issue in the 1976 presidential election campaign and in response, President Ford called a temporary halt to reprocessing to allow time for review a few days before the election. The Carter administration then turned it into a long-term instrument of policy.

a frenzied session of a meeting in Switzerland on this subject. The session was in German, for the benefit of Swiss journalists, and I did not understand much of it, but I kept hearing the word "nonproliferationpolitik" accompanied by expressions of intense anger and banging on the table. At one point Yugoslavia, which purchased a Westinghouse reactor, was close to breaking off diplomatic relations with the United States over this issue.

Not only was withholding fuel shipments a breach of contract, but it was a violation of the International Treaty on Nonproliferation of Nuclear Weapons. That treaty states that a non-weapons nation that signs the treaty is entitled to a secure and uninterrupted supply of fuel for its power reactors. Thus the United States became the first nation to violate that treaty which is the most important safeguard the world has against proliferation. Incidently, this furor in Europe, Asia, and South America over the Carter initiative received virtually no media coverage in this country.

But the worst problem with the Carter initiative was that it failed to achieve much in the way of results. The United States had enough political leverage over South Korea to force that country to cancel its purchase of a reprocessing plant. France canceled its sale to Pakistan, probably in recognition of the fact that that nation had expressed ambitions for building nuclear weapons, but perhaps also partly as a result of American political pressure. However, the German deal with Brazil was not canceled in spite of constant political pressure, including several face-to-face meetings between President Carter and the German Chancellor. The Japanese reprocessing plant was completed and started up. No other reprocessing activity anywhere in the world except in the United States was stopped by it.

While the Carter initiative had little impact on the international proliferation problem, it did have two very important negative effects on this country: it prevented the start-up of the Barnwell plant as discussed earlier, which has had a long-lasting, devastating effect on commercial reprocessing in the United States; and it has completely ruined the U.S. reactor export business. Several nuclear power plants are purchased by foreign countries every year, and at one time, American companies got the lion's share of the business. In recent years, however, the United States has become universally regarded as an unreliable supplier. France and Germany get nearly all of the business. The Soviet Union has become an important international supplier of nuclear fuel.

But perhaps the most disturbing effect of our national nonproliferation politics is that it has caused us to lose most of our influence in international nonproliferation efforts. Before 1977, the United States played leading roles in all international programs and planning to discourage proliferation. We were the leading force in drawing up the International Nonproliferation Treaty and in getting it ratified by most nations of the world. We led the way in seeking and developing technological methods of assuring compliance, and of limiting the problems. However, since the United States "went its own way," we have lost much of our credibility and have had much diminished influence in international nonproliferation programs.

In spite of the general failure of the Carter initiative to achieve its goals and the sizable negative results it did achieve, I have never seen it described in the media as anything but a good and noble program. I have never heard it described as unsuccessful or a failure.

In trying to understand the failure of the Carter effort to stop the spread of reprocessing technology, it is important to consider how effective it might be in stopping weapons proliferation. One obvious limitation was that it was designed only to obstruct the back country road, doing little to obstruct the two main highways to proliferation. Most of the world outside the United States recognized that point and hence regarded the Carter initiative with contempt.

If a nation decides to develop nuclear weapons, lack of reprocessing facilities would hardly stop it. Fourteen nations now have such facilities, and others would have little difficulty in developing them. A commercial reprocessing plant designed to operate efficiently and profitably with minimal environmental impact is a rather expensive and complex technological undertaking, but the same is not true for a plant intended for military use where the only concern is obtaining the product. Construction of such a plant requires no secret information and no unusual skills or experience. Details of reprocessing technology have been described fully in the open literature. It is estimated[13] that a crude facility to produce material for a few bombs could be put together and operated by five people at a cost of $100,000. A plant capable of longer-term production of material for eight bombs per year could be built and operated by 15 people, half of them engineers and the other half technicians, at a cost of $2 million. Either of these plants, or anything in between, could probably be built and operated clandestinely.

Thus stopping reprocessing of commercial power reactor fuel is hardly an effective way of preventing weapons proliferation, and it is not widely viewed as such outside of the United States. On the other hand, reprocessing provides an important source of fuel for present reactors that could tide a needy nation over for a few years in an emergency. It is, furthermore, the key to a future system of breeder reactors which is the only avenue open to many nations for achieving energy independence. Unlike this country, with its abundant supplies of coal, oil, gas, rich uranium ores, shale oil potential, etc., many nations are very poor in energy resources. These include not only heavily industrialized nations like France and Japan, but nations aspiring to that status like Brazil, Argentina, and Taiwan. It is not difficult to understand why these nations are unwilling to trust their very survival to the mercy of Arab sheiks or the whims of American presidents for the indefinite future. They desperately want some degree of energy independence, and reprocessing technology is the key to the only way they can foresee of ever achieving it.

Above and beyond the practical difficulties U.S. nonproliferation policy has encountered, we might ask how important its goal is. There never was any hope that it could prevent a major industrialized nation from developing a nuclear weapons arsenal—there are now five nations with such arsenals. It could only hope to prevent a less developed country like Brazil from taking such a step. By signing the International Nonproliferation treaty, Brazil has renounced any such intentions. The Brazilian reprocessing plant will be subject to very close scrutiny by IAEA inspectors to see that its plutonium is not diverted for use in weapons. Add to this the facts that the plutonium it produces is ill-suited for use in weapons, and that a separate, secret plant could be built and used to produce weapons grade plutonium. It seems clear that the chance is remote that stopping the Brazilian reprocessing plant will be the action that prevents that nation from developing nuclear weapons.

But suppose it did allow Brazil to develop a small arsenal of nuclear weapons—what could it do with it? It could threaten its neighbors, but they could easily be guaranteed against attack by the large nuclear weapons powers; Japan, Germany, and Scandanavia, for example, do not feel threatened by Russian or Chinese nuclear weapons because they are covered by the U.S. umbrella. There are few places in the world where a small nation could use a nuclear bomb these days without paying a devastating price for its action.

If one attempts to develop scenarios that might lead to a major nuclear holocaust, fights over energy resources such as Middle East oil must be at or near the top of the list. Anything that can give all of the major nations secure energy sources must therefore be viewed as a major *deterrent* to nuclear war. Reprocessing of power reactor fuel can provide this energy security, and therefore has an important role in *averting* a nuclear holocaust. That positive role of reprocessing is, to most observers, more important than any negative role it might play in causing such a war through proliferation of nuclear weapons.

After all of this discussion of proliferation, it is important to recognize that use of nuclear power in the United States has no connection to that issue. If we stopped our domestic use of nuclear power, this would not deter a Third World nation from obtaining nuclear weapons, or conversely, use of nuclear power in the United States in no way aids such a nation in obtaining them. The only possible problems occur in transfer of our technology to those countries.

One of the most disturbing aspects of the proliferation problem is the utter lack of information on it that has been made available to the American public through the media. I doubt if more than one percent of the public has any kind of balanced understanding of the subject. Based on the little information provided to them, most people have a distinct impression that our using nuclear power adds substantially to the risk of nuclear war.

This impression has been cemented by the recently developed tactic of antinuclear activists and their media allies to tie nuclear weapons and nuclear power together in one package, purposely making no effort to distinguish between the two. Consider this from an Evans and Novak column after the November 1982 election: "Eight states and the District of Columbia voted for a nuclear freeze (on weapons), but the one crucial issue on any ballot— Maine's referendum on (shutting down) the Yankee Power Plant—the pro-nukes won." The terms "anti-nuke" and "pro-nuke" are more and more being used interchangeably in referring to nuclear weapons and power plants for generating electricity. The propagandists are winning another battle against rationalism.

A TOOL FOR TERRORISTS?

A rather separate issue linking nuclear power with nuclear weapons is the possibility that terrorists might steal plutonium to use for making a bomb. This issue was first brought to public attention in 1973 in a series of articles by John McPhee in the *New Yorker* magazine, later published as a book.[14] He reported on interviews with Dr. Ted Taylor, a former government bomb designer. Taylor had been worried about this problem for some time and had tried to convince government authorities to tighten safeguards on plutonium, which were quite lax at that time, but he could not stir the bureaucracy. The McPhee articles provided an instant solution to the lax safeguards problem— over the next two years, they were dramatically tightened. They also made Ted Taylor an instant hero of the antinuclear movement, and the terrorist bomb issue stayed in the limelight for several years.

Let's take a look at that issue. To begin, consider some of the obstacles faced by terrorists in obtaining and using a nuclear bomb.[15] Their first problem would be to steal at least 20 pounds of plutonium, either from some type of nuclear plant or from a truck transporting it. Any plant handling this material is surrounded by a high-quality security fence, backed up by a variety of electronic surveillance devices, and patrolled by armed guards allowing entry only by authorized personnel. The plutonium itself is kept in a closed-off area inside the plant, again protected by armed guards who allow entry only to people authorized to work with that material. These people must have a security clearance, which means that they are investigated by the FBI for loyalty, emotional stability, personal associations, and other factors that might suggest an affinity for terrorist activities. When they leave the area where plutonium is stored or used, they must pass through a portal equipped to detect the radiation emitted by plutonium. It will readily detect as little as 0.01% of the quantity needed to make a bomb, even if it were in a metal capsule swallowed by the would-be thief. Plants conduct frequent inventories designed to determine if any plutonium is missing. In some plants these inventories are carried on continuously under computer control so as to detect rapidly any unauthorized diversion. There are elaborate contingency plans for a wide variety of scenarios.[16]

When it is transported, plutonium is carried in an armored truck with an armed guard inside. It is followed by an unmarked escort vehicle carrying an armed guard. All guards are expert marksmen qualified periodically by the National Rifle Association.

The truck and the escort vehicles have radio telephones to call for help if attacked, and report in regularly as they travel. There are elaborate plans for countermeasures in the event of a wide variety of problems.[17]

The only significant transport of plutonium in connection with nuclear power would be of ton-size fuel assemblies in which the plutonium is intimately mixed with large quantities of uranium from which it would have to be chemically separated before use in bombs. If terrorists are interested in stealing some plutonium, it would be much more favorable for them to steal it from some aspect of our military weapons program where it is frequently in physical sizes and chemical forms easier to steal and much easier to convert into a bomb. That also would give them *weapons*-grade plutonium, which is much more suitable for bomb making than the reactor grade plutonium from the nuclear power industry. It would be even better for them to steal some high-purity uranium-235 (which is not used in nuclear power activities) from our military program since that is very much easier to make into a bomb. Of course, their best option would be to steal an actual military bomb.

It should be recognized that all of this technology for safeguarding plutonium is now used only for material in the government weapons program. There is essentially no plutonium yet associated with nuclear power. One might wonder how it would be possible to maintain such elaborate security if all of our electricity were derived from breeder reactors fueled by plutonium. The answer is that the quantities of plutonium involved would not be very large. All of the plutonium in a breeder reactor would fit inside a household refrigerator,* and all of the plutonium existing at any one time in the United States would fit into a home living room. The great majority of it would be inside reactors or in spent fuel where the intense radiation would preclude the possibility of a theft. As in the case of radioactive waste, the small quantities involved make very elaborate security measures practical.

There have been charges that all these security measures with armed guards would turn this country into a police state. However, the total number of people required to safeguard plutonium would be only a small fraction of the number now used for security checking in airports to prevent hijacking of airplanes. That force has hardly given our country a police state character.

If the terrorists do manage to steal some plutonium from nuclear power operations and evade the intensive police searches that are certain to follow their theft, their next problem is to fabricate it into a bomb. Ted Taylor's assessment[18] of that task is indicated by the following quote:

> Under conceivable circumstances, a few persons, possibly even one person working alone, who possess about 10 kilograms of plutonium and a substantial amount of chemical high explosive could, within several weeks, design and build a crude fission bomb.
>
> By a "crude fission bomb" we mean one that would have an excellent chance of exploding with the power of at least 100 tons of chemical high explosive . . . The key persons or person would have to be reasonably inventive and adept at using laboratory equipment. . . . They or he would have to be able to understand some of the essential concepts and procedures that are described in widely distributed technical publications concerning nuclear explosives, nuclear reactor technology, and chemical explosions (and) would also have to be willing to take moderate risks of serious injury or death.

Taylor suggested some available publications that would be useful. I read them, but the principal message I derived was that designing a bomb would be even more difficult than I had previously believed it to be. Perhaps some obscure statements in those publications contain the key to solving the problem; they would be readily recognized by a professional government bomb designer, but they were not at all recognizable to me.

---

* A large breeder reactor contains about 3000 kg of plutonium, which, in the form used (plutonium oxide), has a volume of 9 cubic feet. If all our electricity were derived from these reactors, about 400 would be needed, containing 3600 cubic feet of plutonium, the volume of a living room 20 ft $\times$ 18 ft $\times$ 10 feet high.

In order to better understand the difficulties a terrorist would face in fabricating a bomb, let us consider opinions from some other professional government bomb designers. The following statement was obtained from J. Carson Mark of Los Alamos National Laboratory[19]:

> I think that such a device could be designed and built by a group of something like six well-educated people, having competence in as many different fields.
>
> As a possible listing of these, one could consider: A chemist or chemical engineer; a nuclear or theoretical physicist; someone able to formulate and carry out complicated calculations, probably requiring the use of a digital computer, on neutronic and hydrodynamic problems; a person familiar with explosives; similarly for electronics; and a mechanically skilled individual.
>
> Among the above (possibly the chemist or the physicist) should be one able to attend to the practical problems of health physics.
>
> Clearly depending on the breadth of experience and competence of the particular individuals involved, the fields of specialization, and even the number of persons could be varied, so long as areas such as those indicated were covered.

Note that this assessment is more optimistic from the public security viewpoint than Ted Taylor's statement. An even more optimistic assessment was given by E. M. Kinderman of Stanford Research Institute[20]:

> Several people, five to 10 with a hundred thousand dollars or so could do the job if the people were both dedicated to their goal and determined to pursue it over two years or more.
>
> One or a few competent physicist-engineers could probably arrive at a tentative design in a year or so.
>
> A chemical engineer and a metallurgist could . . . construct the essential equipment, make essential tests, and alone or with some help, operate a plant to produce the product dictated by the bomb designer.
>
> Others would be needed for the design and construction of the miscellaneous parts. . . . It is likely that the team will produce something with a force equivalent to 50 to 5000 tons of TNT* . . . and it will weigh less than one ton.

Aside from the three statements quoted, the only other information from bomb experts with which I am familiar was a government statement[21]:

> . . . a dedicated individual could conceivably design a workable device. Building it, of course, is another question and is no easy task.

---

* TNT is the chemical explosive referred to in the statement by Taylor quoted above.

> However, we also recognize that it is conceivable that a group with knowledge and experience in explosives, physics, metallurgy, and with the requisite financial resources and nuclear materials could, over a period of time, perhaps even build a crude nuclear explosive.

The principal information given to the public via the media was based on the separate efforts of an unnamed MIT student, and John Phillips, a Princeton student. The MIT student was hired for a summer to see how far he could get in trying to make a bomb, and the story was told on a "NOVA" television program. He produced a design and fabrication procedure on paper which a Swedish expert judged to have "a small but real chance of exploding with a force of 100–1000 tons [of TNT]." It was not stated whether this referred to a bomb made from weapons-grade or reactor-grade plutonium. It was also not clear what is meant by a Swedish expert, since Sweden has never built a bomb and has always professed disinterest in such an undertaking.

A U.S. government expert who examined the MIT student's fabrication procedure told me that anyone trying to follow it would almost surely be blown up by the chemical explosives long before his bomb was completed. Obviously the student had no experience in handling high explosives. This is an example of the various problems a terrorist would face. There is nothing secret about handling high explosives, but few people are experienced in this type of work. The handling includes cutting and shaping it, and attaching things to it—I have 30 years of experience in experimental physics and have mastered numerous techniques, but I would never consider undertaking anything so dangerous.

The effort of the Princeton student, John Phillips, was much more widely publicized. He made extravagant claims that his bomb would explode with a force of more than 10,000 tons of TNT. He took on a publicity agent, wrote a book, appeared on many TV and radio shows, received very wide newspaper coverage, and even ran for Congress.

What he claimed to have produced was a design for a bomb in a term paper prepared for a physics course. I spoke to his professor in that course, who said that there was nothing in his paper that would ordinarily be called a design. There were only crude sketches without dimensions (i.e., sizes). There were no calculations to support his claim that his bomb would work. He had collected a lot of information that would be useful in designing a bomb, for which the Professor gave him an "A" grade.

His professor told me that he himself had been contacted by many newsmen, but they never printed what he told them—they only trumpeted that Phillips had designed a workable bomb.

Several people have told me that professional government bomb designers have said that a design for a bomb by some student would work. I know that this could not be true because it would be a very serious breach of security regulations for a person who was ever involved with the government program to comment on a design that is available to the public. Note that the MIT student's design was judged by a "Swedish expert." With regard to such claims about Phillips' design, no professional could possibly consider a sketch without dimensions to be a design capable of being evaluated for performance. Science and technology are highly quantitative disciplines, but apparently Phillips—and the media—do not understand that fact.

There have been numerous statements in newspapers, including our University paper, that any college student could design a nuclear bomb. In reply, I published an offer in our University paper of an unqualified "A" grade in both of the two courses on nuclear energy that I teach for any student who can show me a sketch of a workable plutonium bomb together with a quantitative demonstration that it would work. My offer has been repeated every year, for about five years. Two students turned in papers, but neither of them had as much as 5% of what could be called a design.

All of this discussion has been about designs on paper, but as is clear from the above quoted statements by experts, that is only a small part of the task faced by terrorists. The fabrication requires a wide degree of expertise and experience in technical areas. It requires people capable of carrying out complex physics and engineering computations, handling hazardous materials, arranging electronically for a hundred or so triggers to fire simultaneously within less than a millionth of a second, accurately shaping explosive charges, attaching them precisely and connecting the triggers to them, etc. Where would the terrorists find this expertise?

Experienced and talented scientists and technicians generally enjoy well-paid and comfortable positions in our society and hence are not likely to be inclined toward antisocial activity. Recruiting would have to be done under strictest secrecy, which would have to be maintained over the development period of many weeks. Even one unsuccessful recruiting attempt could blow their operation. Moreover, a participant would face a high risk of being killed in his work. And if the plot were discovered he would face imprisonment, not to mention an end to a promising career. The terrorists would surely face severe difficulty in obtaining the needed expertise.

But suppose, somehow, the terrorists succeeded in stealing the plutonium and making the bomb. Let us say it has the explosive force of 300 tons of TNT, which is an average of the various estimates by experts. What could they do with it?

We usually think of a nuclear bomb as something capable of destroying a whole city, but that refers to bombs thousands of times larger. A bomb of this size would roughly be capable of destroying one city block, or one very large building. Ted Taylor uses,[18] as an example, the World Trade Center in New York City, which sometimes contains 50,000 people. That is the origin of the oft-quoted statement that one of these bombs could kill 50,000 people. It could also kill a similar number in a sports stadium by showering them with radiation and burning them with searing heat.

However, if killing 50,000 people is their desire, there are many easier alternatives for accomplishing it. They could

- Release a poison gas into the ventilation system of a large building.
- Dynamite the structural supports in a sports stadium so as to drop the upper tier down on top of the lower tier; this should kill nearly all the people in both tiers.
- Discharge a large load of napalm (or perhaps even gasoline) on the spectators in a sports stadium, either by airplane or by truck.
- Blast open a large dam; there are situations where this could kill over 200,000 people.
- Poison a city water or produce supply.

Any imaginative person could add many more items to this list.

Terrorists with a nuclear bomb would probably first try to use it as an instrument for blackmail. But nonnuclear tactics would be equally useful for that purpose. Unlike the nuclear bomb situation, there are plenty of people with all the know-how to carry out these actions.

Also unlike the nuclear bomb situation, we are doing nothing to avert their implementation, although there are many things we might do. We could guard ventilation systems of large buildings, but we don't. We could guard dams and reservoirs, but we don't. We could inspect sports stadium structural supports for dynamite before major events, but we don't. The high school I attended was recently rebuilt without windows, making its 3000 students defenseless against poison gases introduced into its ventilation system. Terrorists could easily turn theaters or arenas into blazing infernos with blocked exits, but we don't guard against that. They could kidnap wives and children of Congressmen and other high officials, which might be very effective for their purposes, but we show no concern about that problem.

We are vulnerable to mass murder or blackmail by terrorists on dozens of fronts. All of them would be equally effective and infinitely easier for terrorists to take advantage of than making a bomb from plutonium stolen from the nuclear power industry. Yet there is a continuous stream of concern expressed about the latter problem, and none about the others. Once again the antinuclear activists, and the media which dance to their tune, have generated enormous public misunderstanding. They have blown a relatively minor issue grossly out of proportion to its real importance.

For once, however, their activities may have served a useful purpose. Experts on terrorism have told me that it would be very favorable if terrorists devoted their energies to nuclear terrorism and were thus diverted from the easier and more destructive options available to them.

PLUTONIUM TOXICITY

Another property of plutonium unrelated to its use in bombs has attracted a great deal of attention. That is, its toxicity, as exemplified by Ralph Nader's statement that a pound of plutonium could kill 8 billion people.[22] Let's look into that question.

In Chapter 5 we showed how to calculate the toxicity of plutonium ingested into the stomach, which is the way it would most probably enter the human body if it is buried deep underground as part of radioactive waste. However, the most important health effects due to plutonium released from nuclear facilities occur when it becomes suspended in the air as a fine dust and is thereby inhaled into the lungs.*

It is straightforward to quantify the risks associated with this problem. When plutonium oxide, the form in which plutonium would be used in the nuclear industry and also its most toxic form, is inhaled as a fine dust, 25% of it deposits in the lung, 38% deposits in the upper respiratory tract, and the

---

* If this seems contradictory to the discussion in Chapters 5 and 6, the following may be useful:

- for material suspended in air as a fine dust, the probability for it to be inhaled by a human is about one chance in a million,
- for plutonium buried deep underground, the probability that it will be inhaled by a human is about one chance in a trillion, a million times less,
- for plutonium buried deep underground, the chance that it will enter a human stomach is about one chance in 30 million,
- a quantity of plutonium inhaled is about 5000 times more likely to cause cancer than the same quantity taken into the stomach, because in the latter case 99.99% of it is rapidly excreted. For most other materials, the difference between inhalation and dietary intake is very much less.

remainder is exhaled.[23] Within a few hours, all of that deposited in the upper respiratory tract, but only 40% of that deposited in the lung is cleared out. The other 60% of the latter—(0.25 × 60 =)15% of the total inhaled—remains in the lung for a rather long time, an average of two years.

From the quantity of plutonium in the lung and the length of time it stays there, it is straightforward to calculate the radiation exposure to the lung in millirem. For example, a trillionth of a pound gives a dose of 1300 millirem over the two-year period (cf. Appendix). From studies of the Japanese atomic bomb survivors, of miners exposed to radon gas, and other such human exposure experiences, estimates have been developed for the cancer risk per millirem of radiation exposure to the lung.[24,25] Multiplying this by the sum of the radiation doses in millirems received by all those exposed then gives the number of cancers expected. The result is that we may eventually expect about 2 million cancers for each pound of plutonium inhaled by people[26,27] (cf. Appendix).*

There is no direct evidence for plutonium-induced cancer in humans, but there have been a number of experiments on dogs, rabbits, rats, and mice. The results of these are summarized[28] in Fig. 19, where the curve shows the expectation from our calculation. It is evident that the animal data give strong confirmation for the validity of the calculation.

The 2 million fatalities per pound inhaled leaves plutonium dust far from "the most toxic substance known to man." Biological agents, like botulism toxin or anthrax spores[29] are many hundreds or thousands of times more toxic. Plutonium toxicity is similar to that of nerve gas,[29] but given the choice of being in a room with equal quantities of plutonium dust and nerve gas, the latter would be infinitely more dangerous. It rapidly permeates the room air whereas plutonium, being a solid material, would be largely immobile.

In fact, it is rather difficult to disperse plutonium in air as a respirable dust. Individual particles tend to stick together and agglomerate into lumps of too large a size to be inhaled. In the experiments on animals, substantial effort and ingenuity was required to overcome this problem[30] and arrange for the plutonium dust to be inhaled.

The calculational procedure used here to obtain our result, two million deaths per pound inhaled, follows the recommendations of the International Commission on Radiological Protection (ICRP). It would be impossible to obtain a very different result without sharply deviating from them; at least three independent investigations have used them to evaluate the toxicity of

---

* By a more complex process, inhaled plutonium can also cause liver cancer, and to a lesser extent bone cancer. Our treatment here is thus oversimplified.

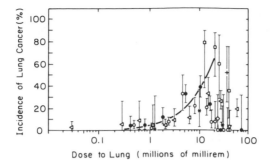

Fig. 19. Results of animal experiments with inhaled plutonium. The curved line shows the predictions of the calculation outlined in the text. Data are compiled in Ref. 28 and the calculation is explained in detail in Ref. 26. □, PuO$_2$, dogs, J.F. Park, private communication to R.C. Thompson quoted in BNWL-SA-4911; ∇, PuO$_2$, mice, L. A. Temple *et al.*, *Nature,* **183,** 498 (1959); △, PuO$_2$, mice, L. A. Temple *et al., ibid;* ◇, PuO$_2$, mice, R. W. Wager *et al.,* Hanford Report HW-41500 (1956); ○, Pu, citrate, rats, L. A. Buldakov and E. R. Lyubchanskii, translation in ANL-tr-864 (1970), p. 381; ●, Pu pentacarbonate, rats, L. A. Buldakov and E. R. Lyubchanskii, *ibid.;* ◁, Pu nitrate, rats, R. A. Erokhim *et al.,* translation in AEC-tr-7387 (1971), p. 344; +, Pu nitrate, rabbits, N. A. Koshurnikova, translation in AEC-tr-7387 (1971), p. 334; *, Pu pentacarbonate, rabbits, N. A. Koshurnikova, *ibid.*

plutonium[26,27] and they have all obtained essentially the same result. These ICRP recommendations are used by all groups charged with setting health standards all over the world, such as the Environmental Protection Agency and the Occupational Safety and Health Agency in the United States. They are almost universally used in the scientific literature.

Nevertheless, there have been at least two challenges to these procedures. The first was based on the so-called "hot-particle" theory, according to which *particles* of plutonium do more damage than if the same amount of plutonium were uniformly distributed over the lung because the few cells near the particles get much larger radiation doses in the former case. The conventional risk estimates are based on assuming that the cancer risk depends on the average dose to all of the cells in the lung, while the hot-particle theory assumes it depends on the dose to the most heavily exposed cells.

This hot-particle theory had been considered by scientists from time to time over the years, but the issue was "brought to head" when the antinuclear activist organization, Natural Resources Defense Council, filed a legal petition asking that the maximum allowable exposures to plutonium be drastically reduced in view of that theory.[31] In response, a number of scientific committees were set up to evaluate the evidence. There were separate committees from the National Academy of Sciences, the National Council for Radiation Protection and Measurements, the British Medical Research Council, and others. All of them independently concluded[32] that there is no merit to the hot-particle theory, and that, if anything, concentration of the plutonium into particles is less dangerous than spreading it uniformly over the lung. The scientific evidence is too complex to review here, but a few points are easily understood:

- Animal experiments with much "hotter" particles of plutonium gave fewer cancers than those with normal particles containing the same total amount of plutonium.
- Particles are constantly moving from place to place, so during their two-year residence in the lung, all cells are more or less equally irradiated.
- There were about 25 workers from Los Alamos National Laboratory who inhaled a considerable amount of plutonium dust during the 1940s; according to the hot-particle theory each of them has a 99.5% chance of being dead from lung cancer by now, but there has not been a single lung cancer among them.[33]

As a result of these studies, the scientific community has rejected the hot-particle theory and standard-setting bodies have not changed their allowable exposures. However, antinuclear activists continue to use the theory to justify such widely quoted remarks as "a single particle of plutonium inhaled into the lung will cause cancer."

Shortly after the fuss over the hot-particle theory had cleared away, John Gofman, head of a San Francisco antinuclear activist organization, proposed[34] a new theory of why plutonium should be much more dangerous than estimated by standard procedures. His theory was based on the idea that the cilia (hairs) that clear foreign material from the bronchial regions are destroyed in cigarette smokers, allowing the plutonium to stay there for years rather than being cleared within hours. He ignored direct experiments[35] which showed unequivocally that dust is cleared from these regions just as rapidly in smokers as in nonsmokers; apparently smokers have more mucous flow and do more coughing to make up the difference. In fact, if smokers cleared dust from their bronchial passage as slowly as Gofman assumes, they would die of suffocation. Gofman made errors in his calculation, as in using a surface area of the bronchi which is 17 times too small. His paper was severely critiqued by a number of scientists.[36,26] It has never gained any acceptance in the scientific community and has been ignored by all committees of experts and standard-setting groups. I know of no scientist other than Gofman who uses it in his work.

When my paper on plutonium toxicity[26] was first published, including its estimate of 2 million cancers per pound of plutonium inhaled, Ralph Nader asked the Nuclear Regulatory Commission to evaluate it. Judging from the number of telephone calls I received asking about calculational details, they did a rather thorough job, and in the end they gave it a "clean bill of health." Nevertheless, Nader continued to state, in his speeches and writings, that a pound of plutonium could kill 8 billion people, 4000 times my estimate. In fact he accused me[37] of "trying to detoxify plutonium with a pen."

In response I offered to inhale publicly many times as much plutonium as he said was lethal. At the same time, I made several other offers for inhaling or eating plutonium—including to inhale 1000 particles of plutonium of any size that can be suspended in air, this in response to "a single particle . . . will cause cancer," or to eat as much plutonium as any prominent nuclear critic will eat or drink caffeine. My offers were such as to give me a risk equivalent to that faced by an American soldier in World War II, according to *my* calculations of plutonium toxicity which followed all generally accepted procedures. These offers were made to all three major TV networks, requesting a few minutes to explain why I was doing it. I feel that I am engaged in a battle for my country's future, and hence should be willing to take as much risk as other soldiers.

None of the TV networks responded (except for a request by CBS for a copy of my paper), so nothing ever came of my offer. However, antinuclear activists have used it to make me seem irrational—they say I offered to eat a *pound* of plutonium, whereas it was actually 800 milligrams, 550 times less. Some people have told me that antinuclear activists get so much media attention because they offer drama and excitement. It seems to me that my offer would have provided these, so there goes another explanation for why the media are so unbalanced on nuclear power issues!

One story in connection with my offer gives insight into why journalists have performed so poorly in informing the public about radiation hazards. A national correspondent for the *Dallas Times-Herald* quoted me as saying "I offered to inhale a thousand times as much plutonium as [Ralph Nader] would eat caffeine." I wrote a letter complaining about this and a host of other errors in his piece, to which his editor replied in part: "[You wrote] 'I offered to inhale a thousand times as much plutonium as he claims would be lethal, and to eat as much plutonium as he would eat caffeine' . . . This seems to be faithful to what our [correspondent] reported."

It is 5000 times more dangerous to inhale plutonium than to eat it, and eating plutonium is about equal in danger to eating the same quantity of caffeine. Thus, if I were to do what the report said I offered to do, I would be taking (1000 × 5000 =) 5 million times greater risk than Nader would be taking in eating the caffeine—I would surely be dead. Actually I offered to *eat* (not inhale) the *same* amount (not 1000 times as much) of plutonium as he would eat caffeine, giving us equal risks. My offer to inhale plutonium was a completely separate item, intended to point out the ridiculousness of his statements about the dangers of inhaling plutonium. How a national correspondent can interpret my quote as he did, and how an editor can then fail to understand the difference when it is pointed out to him, are beyond my comprehension. Nevertheless, it is people like them rather than the scientists who are educating the public about radiation. Note that this is not a question of qualitative versus quantitative; being in error by a factor of 5 million is hardly a matter of lack of precision.

In evaluating the hazards from plutonium toxicity, it gives little insight to say that we can expect 2 million cancers per pound of plutonium inhaled unless we specify how much plutonium would be inhaled in various scenarios. This, of course, depends on the type of release, the wind and other weather conditions, as well as the number of people in the vicinity. But let us say that one pound of plutonium oxide powder is released in the most effective way in an average big-city location under average weather conditions.[26] In the hour or so before the wind blows this dust out of the densely populated areas, only about one part in 100,000 would be inhaled by people,[38] enough eventually to cause 19 cancers. If people know about the plutonium, as in a blackmail situation, they could breathe through a folded handkerchief or piece of clothing, which would reduce the eventual death toll from 19 to 3. Better yet, they could go inside buildings and shut off outside air intakes for this critical short time period.

Eventually all of the plutonium dust would settle down to the ground, but there would still be the possibility of its later being resuspended in air by wind or human activities. Its ability to be resuspended is reduced by rain, dew, and other natural processes, as a result of which the principal threat from this process diminishes rapidly over the first year and essentially disappears thereafter.[39] All in all, this resuspended plutonium dust eventually causes about seven deaths.

Within a few years the plutonium works its way downward into the ground, becoming a permanent part of the top layers of soil. As is well known, it remains radioactive for a very long time. How much harm it does over that period depends on its probability of getting suspended in air by plowing, construction, or natural processes, and then being inhaled by humans. Due to these processes, an average atom of a heavy metal in the top eight inches of soil has 13 chances in a billion of being inhaled by a human each year.[26] If this probability is applied to the plutonium, it will cause a total of only 0.2 additional deaths over the tens of thousands of years that it remains radioactive.

During this period, plutonium in the soil can also be picked up by plant roots, thereby getting into food. This process has been studied in many controlled experiments and in various contamination situations such as bomb test sites and waste disposal areas.[40] Its probability is highly variable with geography, but even under the most unfavorable conditions, this would lead to less than one additional fatality over the tens of thousands of years.

In summary, a pound of plutonium dispersed in a large city in the most effective way would cause an average of 19 deaths due to inhaling from the dust cloud during the first hour or so, with 7 additional deaths due to resuspension during the first year, and perhaps one more death over the remaining tens of thousands of years it remains in the top layers of soil. This gives an ultimate total of 27 eventual fatalities per pound of plutonium dispersed.[26]

It has often been suggested that plutonium dispersion might be used as an instrument for terrorism. But this is hardly realistic because none of the fatalities would occur for at least ten years,* and most would be delayed 20 to 40 years. It could not be used for blackmail because if the dispersal is recognized, protective action is easily taken—breathing through handkerchiefs or going indoors. Terrorists would do much better with nerve gas, which can be made from readily available chemicals; it leaves dead bodies at the scene.

There have been fears expressed that we might contaminate the world with plutonium. However, a simple calculation shows[26] that, even if all the world's electric power were generated by plutonium-fueled reactors, and all of the plutonium ended up in the top layers of soil, it would not nearly double the radioactivity already there from natural sources, adding only a tiny fraction of one percent to the health hazard from that radioactivity. As is evident from the previous discussion, plutonium in the ground is not very dangerous because there is no efficient mechanism for transforming it into airborne dust.

John Gofman, the antinuclear activitist whose work has been discussed previously, has been speaking and writing about effects of plutonium toxicity on the basis of what he calls 99.99% containment.[41] By this he means that 0.01% of all plutonium used each year will be released *as a respirable dust*

---

* Lung cancers resulting from radiation exposure take *at least* ten years to develop, and typically take about 30 years. This is the latent period discussed in Chap. 2.

that will remain suspended in air for long time periods. It turns out that even the steel and asphalt industries do better than that in containing their products,[42] often holding respirable dust releases down to 0.001%. But in plutonium handling, releases are very much smaller. Let's consider the reasons for this.

In the steel and asphalt industries, the materials are heated far above the melting point, resulting in vigorous boiling, which is a prime mechanism for converting some of the material into airborne dust; in the nuclear power industry, plutonium would never be heated to anywhere near its melting temperature. During processing, steel and asphalt are handled in open containers, well exposed to the building atmosphere, whereas plutonium is always tightly enclosed and completely isolated from the building atmosphere. The air from the building atmosphere can mix with outside air only after passing through filters. In steel and asphalt plants these filters consist of ordinary fabric, whereas in plutonium plants they are specially developed high-efficiency filters capable of removing 99.9999% of the dust from air passing through them.

Current Environmental Protection Agency (EPA) regulations require that no more than about one part in a billion of the plutonium handled by a plant escape as airborne dust.[43] All plants now operate in compliance with that regulation. It is 100,000 times less than the releases Gofman has been assuming. If all of the electricity now used in the United States were derived from breeder reactors, the maximum allowable releases would be 0.0007 pounds per year. If all plants were in large cities where we have shown that plutonium releases cause 27 deaths per pound, this would correspond to one fatality every 50 years somewhere in the United States. Since these facilities would not be in cities, the consequences would be considerably lower, much less than one death per century.

Of course the EPA regulations do not cover releases in accidents, and there have been some of these. Two of the most notable were fires in a Rocky Flats, Colorado plant for fabrication of parts for bombs where plutonium is handled in a flammable form (the forms used in the nuclear power industry are not flammable).[44] In the earlier fire in 1957, about one part in 300,000—0.002 ounces—of the plutonium that burned escaped as dust.* After that,

---

* If one part in 100,000 of that was inhaled by humans, the total quantity inhaled was only about (0.002/16 lb ÷ 100,000 =) one billionth of a pound. With 2 million cancers per pound inhaled, the expected number of cancers is 0.002—i.e., there is one chance in 500 for a single cancer to result. This was calculated for a plant in a city; the Rocky Flats plant is not in a city, so the actual number would be about ten times smaller.

many improvements were made, so that in the much larger fire in 1969, only one part in 30 million of the plutonium that burned escaped. Safety analyses indicate that new improvements will considerably reduce even this low figure.

Since only a tiny fraction of all plutonium handled each year would be involved in fires or other accidents, much less than one billionth of the total would be released. Thus accidents are a much lesser source of plutonium in the environment than the routine releases which are covered by the EPA regulations, and the total impact of plutonium toxicity in a full breeder reactor electricity system would be less than one death per century in the United States.*

The most important effects of plutonium toxicity by far are those due to nuclear bombs exploded in the atmosphere. Only about 20% of the plutonium in a bomb is consumed, while the rest is vaporized and floats around in the earth's atmosphere as a fine dust. Over 10,000 pounds of plutonium has been released in that fashion by bomb tests to date,[44] enough to cause about 4000 deaths worldwide. Note that the quantity already dispersed by bomb tests is more than 10 million times larger than the annual releases allowed by EPA regulations from an all-breeder-reactor electric power industry.

I am often asked why such tight regulations are imposed on plutonium releases if they involve so little danger. The answer is that government regulators are driven much less by actual dangers than by public concern. They do pay attention to technological practicalities, and it turns out not to be too difficult to achieve very low releases. Costs are taken into account, but all plutonium handling is now in the military program where cost is not such an important factor. The guiding rule for regulators is that all exposure to radiation should be kept "As Low As Reasonably Achievable" (ALARA), and in the case of plutonium releases the regulation cited corresponds to EPA's judgment of what is ALARA.

The difficulty with this system is that the public and the media interpret very elaborate safety measures as indicators of great potential danger. This increases public concern and perpetuates what has become a vicious cycle involving all aspects of radiation protection—the more we protect, the greater the public concern; and the greater the public concern, the more we protect.

One often hears that in large-scale use of plutonium we will be creating unprecedented quantities of poisonous material. Since plutonium is dangerous principally if inhaled, it should be compared with other materials which are

---

* This does not include releases in reactor accidents which are considered separately in Chap. 3. Plutonium plays a relatively minor role in them.

dangerous to inhale. If all of our electricity were derived from breeder reactors, we would produce enough plutonium each year to kill a half trillion people.* But as has been noted previously in Chapter 5, every year we now produce enough chlorine gas to kill 400 trillion people, enough phosgene to kill 18 trillion, and enough ammonia and hydrogen cyanide to kill 6 trillion with each. It should be noted that these materials are gases which disperse naturally into the air if released, whereas plutonium is a solid which is quite difficult to disperse, even intentionally. Of course, plutonium released into the environment will last far longer than these gases, but recall that the majority of the harm done by plutonium dispersal into the environment is due to inhalation within the first hour or so after it is released. The long-lasting nature of plutonium, therefore, is not an important factor in the comparisons under discussion.

One final point about plutonium toxicity that should be kept in mind is that all its effects on human health that we have been discussing are *theoretical*. There is not direct evidence, or epidemiological evidence, that the toxicity of plutonium has ever caused a human death anywhere in the world.

I have been closely associated professionally with questions of plutonium toxicity for several years, and the one thing that mystifies me is why the antinuclear movement has devoted so much energy to trying to convince the public that it is an important public health hazard. Those with scientific background among them must realize that it is a phony issue. There is nothing in the scientific literature to support their claims. There is nothing scientifically special about plutonium that would make it more toxic than many other radioactive elements. Its long half-life makes it *less* dangerous rather than more dangerous as is often implied; each radioactive atom can shoot off only one salvo of radiation, so, for example, if half of them do so within 25 years, as for a material with a 25 year half-life, there is a thousand times as much radiation per minute as if they spread their emissions over 25,000 years as in the case of plutonium.

No other element has had its behavior so carefully studied, with innumerable animal and plant experiments, copious chemical research, careful observation of exposed humans, environmental monitoring of fallout from bomb tests, etc. Lack of information can therefore hardly be an issue. I can only conclude that the campaign to frighten the public about plutonium toxicity

---

* Each reactor produces about 500 kg of plutonium per year, so the 400 reactors needed would produce $(400 \times 500 =)2 \times 10^5$ kg. Since plutonium could kill 2.3×10^6 people per pound[27] if all of it were inhaled by people, the potential toll is $(2 \times 10^5 \times 2 \times 10^6 \cong)4 \times 10^{11}$, a half trillion.

must be *political* to the core. Considering the fact that plutonium toxicity is a *strictly scientific* question, this is a most reprehensible situation.

I am convinced that the public has "bought" the propaganda about the dangers from plutonium toxicity. Ask a layman and he will probably tell you that plutonium is one of the most toxic substances known to man, and is a terrible threat to our health if it becomes widely used. The media accept this as a fact; plutonium toxicity is no longer treated as an issue worthy of their attention. The deception of the American people on this matter has been essentially complete. Lincoln was wrong when he said "you can't fool all the people all the time."

APPENDIX

Fraction of Cost of Nuclear Power Due to Raw Fuel

Nuclear fuel undergoing fission produces 33×10^6 kW-days[45] of heat energy per metric ton (2200 lb). It requires about 6 tons of uranium to make one ton of fuel,* and one-third of the heat energy is converted into electricity.[45] Therefore, the electrical energy per pound of uranium is

$$\frac{1/3 \times 33 \times 10^6 \text{ kW-days}}{6 \times 2200 \text{ lb}} \times \frac{24 \text{ hours}}{\text{day}} = 20,000 \frac{\text{kW-hr}}{\text{lb}}$$

Nuclear electricity costs about 2.5¢/kW-hr, so the electricity production from one pound of uranium costs $(0.025 \times 20,000 =)$ \$500. Uranium costs about \$25/lb, which is 5% of this cost.

Cost of Gasoline versus Cost of Nuclear Fuel

From straightforward energy conversions, 1 lb of nuclear fuel undergoing fission is equivalent to 2.5×10^5 gallons of gasoline. Since the energy conversion efficiency in a breeder reactor[45] is 40%, the electrical energy from 1 lb of uranium, which costs about \$25, is equal to that in 1.0×10^5 gallons.

---

* In isotopic enrichment, 1 lb of natural uranium input contains 0.007 lb U-235, while the depleted product contains 0.002 lb; thus it contributes 0.005 lb of U-235 to the enriched product. One pound of fuel contains 0.03 lb U-235 from this enriched product; it therefore requires $(0.03 \div 0.005 =)$ 6 lb of uranium input.

The equivalent cost of gasoline is therefore $25/1 $\times$ 10^5 gallon, = 0.025¢/ gallon. This is 40 gallons for a penny.

Present reactors burn only 1% of the uranium and are only 33% efficient, so the fuel cost is higher by a factor of 100 $\times$ (40/33) = 120. This is equivalent to gasoline costing (120 $\times$ 0.025 =) 3¢/gallon.

Radiation Dose to Lung from Plutonium and the Lung Cancer Risk

This calculation requires knowing (or accepting) some scientific definitions and may therefore not be understandable to many readers.

We calculate the dose to the lung from a trillionth of a pound of plutonium residing there for two years. The number of plutonium atoms is

$$10^{-12} \text{ lb} \times 450 \text{ g/lb} \times (6 \times 10^{23} \text{ atoms/239 g}) = 1.1 \times 10^{12}$$

where 6 $\times$ 10^{23} is the Avogadro number and 239 is the atomic weight.[45] Since half of the plutonium atoms will decay in 24,000 years (the half-life), the fraction undergoing decay during the two years it spends in the lung is a little more than one in 24,000; actually it is 1/17,000. The number that decay is then 1.1 $\times$ 10^{12}/17,000 = 7 $\times$ 10^7. Each decay releases an energy of about 5 MeV $\times$ 1.6 $\times$ 10^{-13} J/MeV = 8 $\times$ 10^{-13} J, so the total radiation energy deposited is 7 $\times$ 10^7 $\times$ 8 $\times$ 10^{-13} = 5.6 $\times$ 10^{-5} J.[45] The weight of the average person's lung is 0.57 kg[46]; thus the energy deposited is (5.6 $\times$ 10^{-5} ÷ 0.57 =) 1 $\times$ 10^{-4} J/kg. The definition of a millirad[45] is 1 $\times$ 10^{-5} J of energy deposit per kilogram of tissue.* The dose is therefore (1 $\times$ 10^{-4} ÷ 1 $\times$ 10^{-5} =) 10 mrad. Since only 15% of what is inhaled spends this two years in the lung,[23] the exposure per trillionth of a pound *inhaled* is (10 $\times$ 0.15 =) 1.5 mrad. For alpha particles, the radiation emitted by plutonium, 1 mrad = 20 mrem,[47] so the dose to the lung is (1.5 $\times$ 20 =) 30 mrem per trillionth of a pound inhaled.

Estimates by BEIR,[24] UNSCEAR,[25] and ICRP[47] give a risk of about 5 $\times$ 10^{-7} lung cancers per millirad of alpha particle exposure. The number of lung cancers per pound inhaled is therefore (1.5 $\times$ 10^{12} $\times$ 5 $\times$ 10^{-7} =) 8 $\times$ 10^5. Mays[27] estimates 4 $\times$ 10^5 liver and bone cancers per pound inhaled, bringing the total effect to 1.2 million cancers of all types per pound inhaled.

---

* Millirad is the unit of physical radiation exposure, as indicated by this definition, whereas millirem (mrem) includes a correction for biological effectiveness. For X-rays, beta rays, and gamma rays, 1 millirad is equal to 1 millirem, but for alpha particles, 1 millirad is equal to 20 millirem.

There are two factors modifying this estimate. One is that this calculation is for young adults; averaging over all ages reduces the risk in half. The other factor is due to the fact that our calculation was for plutonium-239 whereas typical samples of plutonium contain a mixture of other plutonium isotopes which generally contain more radioactivity per pound because they have shorter half lives.

When these factors are taken into account,[26] the deaths per pound inhaled become 4.2 million for wastes from present reactors, 2.7 million for breeder reactor fuel, and 0.8 million for weapons plutonium. In this chapter we have used 2 million deaths per pound as a loose average, mainly because that number has been used in most studies whose results are quoted here. All results quoted can be adjusted by taking effects to be proportional to these numbers.

REFERENCE NOTES

1. F. R. Best and M. J. Driscoll, *Transactions of the American Nuclear Society,* **34,** 380 (1980).
2. B. L. Cohen, "Breeder Reactors—A Renewable Energy Source," *American Journal of Physics,* **51,** 75 (1983).
3. R. Avery and H. A. Bethe, "Breeder Reactors: The Next Generation," in *Nuclear Power: Both Sides,* M. Kaku and J. Trainer (eds.) (Norton, New York, 1982).
4. T. G. Ayers, *et al.,* LMFBR Program Review, U.S. Energy Research and Development Administration (1978); Report to the Task Forces to the LMFBR Review Steering Committee, Energy Research and Development Administration, April 6 (1977).
5. Joint Committee on Atomic Energy, Review of National Breeder Reactor Program, January (1976).
6. R. Wilson, "Report on the Safety of a Liquid Metal Fast Breeder Reactor," Electric Power Research Institute, Palo Alto, California (1976).
7. J. B. Yasinsky (ed.), "Position Papers on Major Issues Associated with the Liquid Metal Fast Breeder Reactor," Westinghouse Electric Corp., Madison, Pennsylvania (1978).
8. Milton R. Benjamin, *Washington Post,* July 20 (1982).
9. Atomic-Industrial Forum, "Light Water Reactor Fuel Cycles—An Economic Comparison of the Recycle and Throw-away Alternatives," February (1981). This presents analyses by six different groups.
10. Robert Lesch, "World Reprocessing Facilities," *Worldwide Nuclear Power,* January (1982).
11. Shelby T. Brewer, "Letter to Recipients of Worldwide Nuclear Power," dated March 4 (1982).

12. W. Meyer, S. K. Loyalka, W. E. Nelson, and R. W. Williams, "The Homemade Bomb Syndrome," *Nuclear Safety,* **18,** 427 (1977).
13. C. Starr and E. Zebroski, "Nuclear Power and Weapons Proliferation," American Power Conference, April (1977).
14. J. McPhee, *The Curve of Binding Energy,* (Ballantine Books, New York, 1975).
15. U.S. Nuclear Regulatory Commission, "Safeguarding a Domestic Mixed Oxide Industry Against a Hypothetical Subnational Threat," NUREG-0414, May (1978).
16. U.S. Nuclear Regulatory Commission, "Regulatory Guide 5.55: Standard Format and Content of Safeguards Contingency Plans for Fuel Cycle Facilities" (1978); also "Regulatory Guide 5.54: Standard Format and Content of Safeguards Contingency Plans for Nuclear Power Plants" (1978).
17. U.S. Nuclear Regulatory Commission, "Regulatory Guide 5.56: Standard Format and Content of Safeguards Contingency Plans for Transportation" (1978).
18. M. Willrich and T. B. Taylor, *Nuclear Theft: Risks and Safeguards* (Ballinger Publ. Co., Cambridge, Massachusetts, (1974).
19. F. H. Schmidt and D. Bodansky, *The Energy Controversy: The Fight over Nuclear Power* (Albion Press, San Francisco, 1976).
20. B. L. Cohen, "Plutonium—How Great is the Terrorist Threat," *Nuclear Engineering International,* February (1977). The quote from Kinderman is given there. It was taken from a book, but I cannot recall the name of the latter.
21. U.S. Energy Research and Development Administration (ERDA) Safeguards Program, Background Statement," March 10 (1975).
22. R. Nader, speech at Lafayette College, Spring (1975).
23. International Commission on Radiological Protection (ICRP), Task Group on Lung Dynamics, "Deposition and Retention Models for Internal Dosimetry of the Human Respiratory Tract," *Health Physics,* **12,** 173 (1966).
24. National Academy of Sciences Committee on Biological Effects of Ionizing Radiation (BEIR), "Effects on Populations of Exposure to Low Levels of Ionizing Radiation" (1980).
25. United Nations Scientific Committee on Effects of Atomic Radiation (UNSCEAR), "Sources and Effects of Ionizing Radiation" (1977).
26. B. L. Cohen, "Hazards from Plutonium Toxicity," *Health Physics,* **32,** 359 (1977).
27. The Medical Research Council, *The Toxicity of Plutonium* (Her Majesty's Stationery Office, London, 1975); C. W. Mays, "Discussion of Plutonium Toxicity," in R. G. Sachs (ed.), *National Energy Issues—How Do We Decide* (Ballinger Publ. Co., Cambridge, Massachusetts, 1980); C. W. Mays, "Risk Estimates for Liver," in "Critical Issues in Setting Radiation Dose Limits," NCRP (1981).
28. W. J. Bair, "Toxicology of Plutonium," *Advances in Radiation Biology,* **4,** 225 (1974).
29. J. H. Rothchild, *Tomorrow's Weapons* (McGraw-Hill, New York, 1964).
30. J. F. Park, W. J. Bair, and R. H. Busch, "Progress in Beagle Dog Studies with Transuranium Elements at Battelle-Northwest," *Health Physics,* **22,** 803 (1972).

31. A. R. Tamplin and T. B. Cochran, "Radiation Standards for Hot Particles," Natural Resources Defense Council Report, 1974. Also, "Petition to Amend Radiation Protection Standards as They Apply to Hot Particles," submitted to EPA and AEC, Feb. 1974.

32. National Academy of Sciences, "Health Effects of Alpha Emitting Particles in the Respiratory Tract," Environmental Protection Agency Document EPA 520/4-76-013 (1976); National Council on Radiation Protection and Measurements (NCRP), "Alpha Emitting Particles in Lungs," NCRP Report No. 46 (1975); United Kingdom National Radiological Protection Board, Report R-29 and Bulletin No. 8 (1974); W. J. Bair, C. R. Richmond, and B. W. Wachholz, "A Radiobiological Assessment of the Spatial Distribution of Dose from Plutonium," U.S. Atomic Energy Commission Report WASH-1320 (1974); See also The Medical Research Council, Ref. 27.

33. G. L. Voelz, "What We Have Learned About Plutonium from Human Data," *Health Physics*, **29**, 551 (1975).

34. J. W. Gofman, "The Cancer Hazard from Inhaled Plutonium," Committee for Nuclear Responsibility Report CNR 1975-1, reprinted in Congressional Record—Senate 31, July (1975), p. 14610.

35. R. W. Albert *et al.*, *Archives of Environmental Health*, **18**, 738 (1969); **30**, 361 (1975).

36. W. J. Bair, "Review of Reports by J. W. Gofman on Inhaled Plutonium," Battelle Northwest Lab. Report BNWL-2067; C. R. Richmond, "Review of John W. Gofman's Report on Health Hazards from Inhaled Plutonium," Oak Ridge National Laboratory, Report ORNL-TM-5257 (1975); J. W. Healy *et al.*, "A Brief Review of the Plutonium Lung Cancer Estimates by John W. Gofman," Los Alamos Scientific Laboratory, Report LA-UR-75-1779 (1975); M. B. Snipes *et al.*, "Review of John Gofman's Papers on Lung Cancer Hazard from Inhaled Plutonium," Lovelace Foundation, Albuquerque, New Mexico, Report LF-51 UC-48 (1975); "Comments Prepared by D. Grahn," Argonne National Laboratory (1975).

37. R. Nader, *Family Health*, Jan. (1977), p. 53.

38. U.S. Atomic Energy Commission, "Meterology and Atomic Energy," pp. 97ff (1968). This gives the calculational procedures used in Ref. 26.

39. K. Stewart, in *The Resuspension of Particulate Material from Surfaces*, B. R. Fish (ed.) (Pergamon Press, New York, 1964); W. H. Langham, Los Alamos Scientific Lab Report LA-4756 (1971); L. R. Anspaugh, P. L. Phelps, N. C. Kennedy, and H. C. Booth, Proceedings of the Conference on Environmental Behavior of Radionuclides Released in the Nuclear Industry, International Atomic Energy Agency, Vienna (1975).

40. H. R. McLendon *et al.*, International Atomic Energy Agency Document IAEA-SM-199/85, p. 347 (1976)—Savannah River Plant; R. C. Dahlman, E. A. Bondietti, and L. D. Eyman, Oak Ridge National Laboratory, Environmental Sciences Division Publication 870 (1976)—Oak Ridge; F. W. Wicker, Colorado State

University Report COO-1156-80 (1975)—Rocky Flats, Colorado; E. M. Romney, A. Wallace, R. O. Gilbert, and J. E. Kinnear, International Atomic Energy Agency Document IAEA-SM-199/73, p. 479 (1976)—Eniwetok.

41. J. Gofman, *National Forum*, Summer (1979).
42. B. L. Cohen, "Plutonium Containment," *Health Physics*, **40**, 76 (1981).
43. U.S. Environmental Protection Agency, *Federal Register*, **40**, 23420 (1975).
44. U.S. Atomic Energy Agency, "Plutonium and Other Transuranic Elements: Sources, Environmental Distribution, and Biomedical Effects," Document WASH-1359 (1974).
45. J. R. Lamarsh, *Introduction to Nuclear Engineering*, (Addison-Wesley, Reading, Massachusetts, 1975).
46. International Commission on Radiological Protection (ICRP), *Report of the Task Group on Reference Man*, ICRP Publication 23 (Pergamon Press, New York, 1975).
47. International Commission on Radiological Protection (ICRP), *Recommendations of ICRP*, ICRP Publication 26 (Pergamon Press, New York, 1977).

Chapter 8 / COSTS OF NUCLEAR POWER— THE ACHILLES' HEEL*

An important strategy of antinuclear activists has been to stop nuclear power by driving the cost up to the point where it becomes uneconomical. It is difficult not to admire the proficiency with which they have succeeded in attaining their goal. The last order for a nuclear power plant in the United States was placed in 1978, and several dozen partially completed plants have been abandoned since that time, at a cost to the American public of many billions of dollars. But the real *coup de grace* by the antinuclear activists has

* Essentially all of the subjects that have been discussed up to this point in the book have been in areas where I have professional expertise. I have published research papers on them and have a reasonable familiarity with their scientific literature. However, this chapter and the next deal with subjects—economics and solar electricity—not in that category. I have frequently read about them over the years, and I have lectured on them in courses I teach, but I have never published research papers in those fields and cannot claim professional expertise in them. Nevertheless, it is important that they be covered in this book, and I have exerted a rather strenuous effort to treat them competently, including studying over several dozen documents and personally contacting several dozen experts in those fields. What follows, therefore, is the viewpoint of an educated nonexpert.

been in placing the blame for this fiasco on the shoulders of the nuclear power industry, using this argument to cement their victory.

Several large nuclear power plants were completed in the early 1970s at a typical cost of $170 million, whereas plants of that size completed in 1983 cost an average of about $1.7 billion, a tenfold increase. Inflation, of course, has played a role, but it accounts for less than a third of that increase. What has caused the remaining large increase? Ask any antinuclear activist and he will tell you that it is bungling by the nuclear industry. He will recite a succession of horror stories about mistakes, inefficiency, sloppiness, and ineptitude. He will create the impression that people who build nuclear plants are a bunch of bungling incompetents. The only thing he doesn't explain is how these same bungling incompetents managed to build power plants so efficiently, so rapidly, and so cheaply in the early 1970s.

For example,[1] Consumer's Power Company of Michigan completed one of the early nuclear plants, Big Rock Point, in 1962 at a total cost of $27 million. It took $2\frac{1}{2}$ years to construct and has operated very efficiently since that time. In 1976–1977 it ran 343 consecutive days, a world record at that time—the average time between forced shutdowns has been 78 days for nuclear plants and 30 days for coal-burning plants.[2] In 1971, the same company completed its Palisades Plant, with 12 times larger electrical output, at a total cost of $186 million, including $28 million due to licensing delays caused by participation in hearings by antinuclear activists. The total project took 4 years, but the actual construction was completed in 31 months, a national record for plants of that size. The early operation was plagued by faulty equipment, for which Consumers Power collected about $70 million in damages from suppliers, but since 1976, its operation has been good. In 1977, it had the best operating record for large nuclear plants in the United States.

This has hardly been a record of bungling incompetence. But in 1969 Consumer's Power filed an application for a construction permit to build its Midland plant* with only 15% larger electrical output than Palisades. As a consequence of continual legal harrassment by antinuclear activists, who were by that time very well organized, the construction permit was not issued until nearly four years later, and for the following two years construction was halted several times because of legal intervention. The legal battles over the construction permit continued for several more years until finally in 1978 the

---

* There are two Midland plants being built simultaneously, but one of these is partly to provide steam for industrial use. To avoid this complication, our discussion refers only to the other plant.

U.S. Supreme Court ruled unanimously that it was valid, chastising a lower court for "obstructionism." As of now, the plant is scheduled to be completed in 1984, eleven years after start of construction, at a cost of about $1.4 billion, $7\frac{1}{2}$ times the cost of the Palisades plant.

Judging by the success of its earlier plants, it is difficult to blame the high cost of the latest plant on incompetence of the constructors. Moreover, there are many similar examples for other utilities.* The cost of nuclear power plants has indeed skyrocketed, and our purpose here is to understand how and why. We will proceed in short steps.

UNDERSTANDING POWER PLANT CONSTRUCTION COSTS

The Philadelphia office of United Engineers, under contract with U.S. Government agencies, makes frequent detailed estimates[3] of what a large (one million kilowatt) nuclear power plant should cost if constructed with reasonable efficiency and in compliance with all regulations in force at the time. We will use their estimates in the first part of our discussion. The total costs they have estimated are shown in Fig. 20 plotted against the date the project was initiated which is when the estimates were made. The skyrocketing is clearly evident.

The cost of a nuclear plant is defined as the total amount of money spent up to the time it goes into operation. This is the product of three factors:

(1) The cost if all materials and labor were purchased at the time the project is started. This is what is called the EEDB (energy economics data base). Its estimate by United Engineers is a very large undertaking, requiring a listing of the thousands of systems, subsystems, construction materials, pieces of equipment, categories of labor, engineering projects, etc., etc., that go into a plant, and estimating the cost of each on a national average. This EEDB, as calculated at various times, is plotted in Fig. 21.

---

* Philadelphia Electric completed its two Peach Bottom plants in 1974 at a cost of about $763 million, but its two Limerick plants of the same size are scheduled for completion in 1985 at a cost of $4.7 billion, 6.2 times higher. Duquesne Light Co. set out to build two *identical* plants; Beaver Valley-1, completed in 1977 cost $611 million, while Beaver Valley-2 is scheduled for completion in 1986 at a cost of $2300 million. Niagara Mohawk completed its Nine Mile Point-1 unit in 1969 for $163 million, but its Nine Mile Point-2 plant, 1.7 times larger, is to be completed in 1986 at a cost of $3.7 billion, 23 times higher. Tennessee Valley Authority completed plants of about the same size in 1974–1977 for $295 million, in 1981–1982 for $865 million, and in 1984–1985 (projected) for $1570 million. Many further examples could be given.

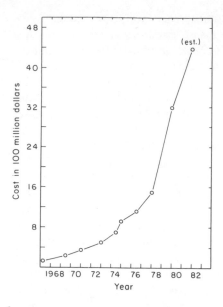

Fig. 20. Cost of 1 million kW nuclear power plants as predicted by United Engineers on the dates shown for plants started at that date. From Ref. 3.

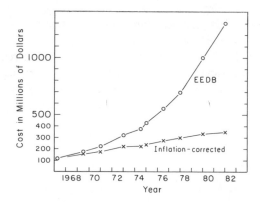

Fig. 21. EEDB cost of 1 million kW nuclear power plants as estimated by United Engineers for the dates shown. This is the cost if all materials and labor were purchased on that date, with no allowance for inflation or interest charges after that date. The lower curve corrects this for inflation by dividing the upper curve by the inflation factor since 1967 according to the Consumer Price Index plus 2% per year escalation of construction labor and materials costs. From Ref. 3.

(2) The cost escalation factor (ESC), which takes into account the inflation of costs with time after project initiation. For each item, this depends on how far in the future it must be purchased: the basic engineering, for example, will be done shortly after the project begins and hence its cost is little affected by inflation, but an instrument that can be installed rapidly and is not needed until the plant is ready to operate may not be purchased for ten years. If the assumed inflation rate is 12% per year, its cost will have tripled by that time ($1.12^{10} = 3$).

(3) A factor covering the interest charges on funds used during construction, INT (which is mathematically related to what is commonly called AFDC, allowance for funds used during construction). All money used for construction must be borrowed (or obtained by some roughly equivalent procedure); hence the interest that must be paid on it up to the time the plant goes into operation is included in the total cost of the plant. For example, the basic engineering may have to be paid for 12 years before the plant becomes operational; if the annual interest rate is 15%, its cost is therefore multiplied by ($1.15^{12} =$)5. Note that the interest which increases item (3) is normally a few points higher than the inflation rate which increases item (2); it therefore pays to delay money outlays as long as possible.

Items (2) and (3), ESC and INT, depend almost exclusively on two things, the length of time required for construction, and the rate of inflation (interest rates, averaged over long periods are closely tied to inflation), increasing rapidly as these increase. If there were no inflation, or if plants could be built very rapidly, these factors would be close to 1.0, having no impact on the cost of the plant.

The product ESC × INT used in the United Engineers estimates at various project initiation dates is plotted in Fig. 22. The number of years

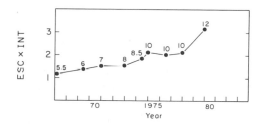

Fig. 22. The product of the inflation and interest factors. This is the factor by which the EEDB from Fig. 2 must be multiplied to obtain the total cost in Fig. 3. The figures above the points are the estimated number of years for the project at its initiation date. From Ref. 3.

required for construction is given above each point. We see that ESC × INT was only 1.17 in 1967 when construction times were 5.5 years and the inflation rate was 4% per year. It increased to 1.45 in 1973 when construction times stretched to 8 years but inflation rates were still only 4%/yr, went up to about 2.1 in 1975–1978 when construction times lengthened to 10 years and the inflation rate averaged about 7% per year, and jumped to 3.2 in 1980 when construction times reached 12 years and the inflation rate soared to 12% per year. That is, the cost of a plant started in 1980 is *more than triple* the expenditures for materials and labor; 69% of the final cost is in inflation and interest.

The inflation rates quoted here are those of the Consumer Price Index (CPI) which is widely used as an estimator for national inflation; e.g., social security payments and many union wage scales are tied to it. The CPI is given for the project initiation dates, but it really should be the average inflation rate over the construction period. In making estimates of the type used here, there is little choice but to assume that inflation will continue at the rate current when the estimate is made (or something similar). But it should be recognized that if inflation were to stop, the cost of a nuclear power plant started now would be less than half of what is estimated in Fig. 20. On the other hand, plants started in the early 1970s cost much more than the original estimates because inflation rates have been a lot higher than was expected at the time the original estimates were made. Similar statements apply to the time required for construction. If this should be reduced from its present 12 years back to something like six years, the cost of a plant would be cut about in half even if inflation continued unchanged. On the other hand, the cost of plants being completed now has been greatly increased by the fact that their construction times have been much longer than originally planned.

The total cost of a plant, shown in Fig. 20, is the product of the (ESC × INT) shown in Fig. 22, and the EEDB shown in Fig. 21. We not turn to a discussion of the latter, the cost if all materials and labor were purchased and paid for on the project initiation date. This is, of course, increased by inflation occurring up to that date. We might think that its increase follows the Consumer Price Index, but costs of construction labor and materials have been increasing more rapidly than general inflation as measured by the CPI by about 2% per year[4]; for instance,[5] the average annual price increase between 1973 and 1981 was 11.5% for concrete, 10.2% for turbines, and 13.7% for pipes, versus only 9.5% for the CPI. When this additional 2% per year inflation is taken into account, the inflation corrected cost estimates are shown by the line so labeled in Fig. 21. It indicates a doubling between 1971 and 1982; that is, the quantity of labor and materials used in constructing a plant doubled during that time period.

REGULATORY RATCHETING

It is important to understand this doubling in the amount of labor and materials going into a nuclear power plant between 1971 and 1982. The Nuclear Regulatory Commission (NRC) and its predecessor, the Atomic Energy Commission Office of Regulation, as parts of the United States Government, must be responsive to public concern. Since the early 1970s there has been mounting public concern about the safety of nuclear power plants; NRC has therefore responded in the only way it can, by tightening regulations and requirements for safety equipment.

Make no mistake about it, you can always try to improve safety by spending more money . Even with our personal automobiles, there is no end to what we can spend for safety—larger and heavier cars, blow-out-proof tires, air bags, passive safety restraints, rear window wipers and defrosters, fog lights, more shock-absorbent bumpers, and so on. In our homes we can spend large sums on fireproofing, sprinkler systems, and fire alarms, to cite only the fire protection aspect of household safety. Nuclear power plants are much more complex than homes or automobiles, leaving innumerable options for spending money to improve safety. In response to escalating public concern, NRC has been implementing some of these options.

This process has come to be known as "ratcheting." Like a ratchet wrench that is moved back and forth but always tightens and never loosens a bolt, the regulatory requirements have been constantly tightened, requiring additional equipment and construction labor and materials. Between the early and late 1970s, the poundage of steel needed in a power plant of equivalent electrical output increased by 41%, the amount of concrete by 27%, the lineal footage of piping by 50% and of electrical cable by 36%, and the man-hours of labor by 50% according to one study.[6] No one can recall a situation where NRC withdrew a requirement made in the early days on the basis of minimal experience because later experience demonstrated that it was unnecessarily stringent. The ratcheting policy was consistently lived up to.

Commonwealth Edison, the utility which serves the Chicago area, completed its Zion plants in 1973–1974 with 5.3 man-hours of labor per kilowatt of plant output (i.e., 5.3 million man-hours for a one million kilowatt plant); its Byron plants, scheduled for completion in 1984–1985 will use an estimated 12.8 man-hours per kilowatt.[2] For plants started now, United Engineers estimates 23 man-hours per kilowatt[3] would be required.

In its regulatory ratcheting activities, NRC has always paid some attention to cost effectiveness, attempting to balance safety benefits against cost increases. However, the NRC personnel privately concede that their cost estimates were very crude, and often unrealistically low. (NRC has recently set up a Cost Estimation Group to improve on that situation.) Estimating costs of tasks never before undertaken is, at best, a difficult and inexact art. The United Engineers EEDB for 1980 estimated 17 man-hours per kilowatt whereas their EEDB for 1982 estimated 23 man-hours per kilowatt; the difference was less due to new regulatory requirements[3] than to field experience which showed that some of the requirements instituted earlier were more difficult to satisfy than had been anticipated at the time of the 1980 estimates.

In addition to increasing the quantity of materials and labor going into a plant, regulatory ratcheting has increased costs by extending the time required for construction. According to the United Engineers estimates, the time from project initiation to ground breaking[7] was 16 months in 1967, 32 months in 1972, and 54 months in 1980. These are the periods needed to do initial engineering and design, to develop a safety analysis and an environmental impact analysis supported by field data, to have them reviewed by the NRC staff and its Advisory Committee on Reactor safeguards and work out conflicts with these groups, to subject them to criticism in public hearings and to respond to that criticism (often with design changes), and finally, to receive a construction permit. The time from ground breaking to operation testing was increased from 42 months in 1967, to 54 months in 1972, to 70 months in 1980.

According to Fig. 20, the cost of a plant started in 1982 is 12 times the cost of one started in 1971. Roughly, this factor of 12 consists of a factor of 2 due to increased labor and materials, a factor of 2 due to lengthened construction time, and a factor of 3 due to inflation—$2 \times 2 \times 3 = 12$. Regulatory ratcheting, which is responsible for the first two of these factors, has therefore quadrupled the real cost of a nuclear power plant. A plant started in 1982 shown in Fig. 21 as costing $4.3 billion when completed in 1994 could have been completed in 1989 at a cost of $1 billion if there had been no regulatory ratcheting.

What has all this regulatory ratcheting bought in the way of safety? One point of view often expressed privately is that it has bought *nothing*. A nuclear power plant is a very complex system and adding to its complexity involves a risk in its own right. If there are more pipes, there are more ways to have pipe breaks, which are one of the most dangerous failures in reactors. With more complexity in electrical wiring, there is more chance for a short circuit, or for an error in hook-ups, and less chance for such an error to be discovered. On the other hand, each new safety measure is aimed at reducing a particular safety shortcoming and undoubtedly does achieve that limited objective. It would be impossible to prove or disprove the thesis that the added complexity reduces safety more than attaining the limited objective increases safety.

A more practical question is whether the escalation in regulatory requirements is necessary, justified, or cost effective. The answer depends heavily on one's definition of those words. The nuclear regulators of 1967 and 1973 were quite satisfied that plants completed and licensed at those times were adequately safe, and the great majority of knowledgeable scientists agreed with them. With the exception of improvements instigated by lessons learned in the Three Mile Island accident, which increased the cost by only a few percent, there have been no new technical developments indicating that more expenditures for safety are needed. In fact, the new developments suggest the contrary (cf. Chapter 3). The most significant result of safety research during the last decade has been finding that the emergency core cooling system works better than expected and far better than indicated by pessimistic estimates from the 1973 time period. The second most important result was finding that radioactive iodine in a water environment behaves much more favorably than had been assumed. There has been no new information indicating that low-level radiation is more dangerous than estimated in 1973; the risk estimates of the National Academy of Sciences BIER Committee have been lowered somewhat, and no official or prestigious committee has raised its estimates of the danger. The most unfavorable development that might be interpreted as impacting on safety, increased concern about pressurized thermal shock (cf. Chapter 3), has an impact on maintenance costs but not on initial construction costs. Indeed, it is not an issue at all for reactor vessels fabricated since 1971.

Clearly, the regulatory ratcheting has been driven not by new scientific or technological information, but by public concern and the political pressure it generates. Of course changing regulations as new information becomes available is a normal process, but it would normally work both ways. The ratcheting effect, only making changes in one direction, is an abnormal aspect of recent regulatory practice which is completely unjustified from the scientific point of view. It has been a strictly *political* phenomenon that has quadrupled the cost of nuclear power plants.

ACTUAL COSTS OF NUCLEAR POWER PLANTS—REGULATORY TURBULENCE

All of our discussion to this point has been based on United Engineers estimates of costs for a nuclear power plant built with reasonable efficiency in accordance with all regulations in force on the project initiation date. Unfortunately, actual nuclear power plants are not built that way. As new regulations are promulgated, plants under construction must be modified to incorporate them. We refer to effects of these regulatory changes made during the course of construction as "regulatory turbulence," and the reason for that name will soon become evident.

As anyone who has tried to make major changes in design of his house while it was under construction can testify, doing so is a very time-consuming and expensive practice, much more expensive than including the changes in the original design. There have been situations where the walls of a building were already in place when new regulations appeared requiring substantial amounts of new equipment to be included inside them[8]—in some cases this could be nearly impossible, and in most cases it requires a great deal of extra expense for engineering and repositioning of equipment, piping, and cables that have already been installed. In some cases it may require chipping out concrete that has already been poured, an extremely expensive proposition.

In attempting to avoid such situations, constructors often design features into a power plant that are not required in anticipation of rule changes that may never materialize. This also adds to the cost.

Frequently changing plans in the course of construction is a confusing process which can easily generate mistakes, and mistakes can be costly. The Diablo Canyon plant in California was ready to operate when such a mistake was discovered, necessitating many months of delay. For a $1.5 billion plant built with money borrowed at 16% interest, delay in start-up costs ($1.5 billion $\times$ 0.16 =) $240 million per year, or $650,000 per day.

Since delay is so expensive, plant constructors often choose to do things that appear to be very wasteful. Construction labor strikes must be avoided at almost any cost. There was a well-publicized situation on Long Island where a load of pipe delivered from a manufacturer did not meet size specifications; rather than return it and lose time, the pipe was machined to specifications on site, at great added expense.

A very important source of cost escalation in some plants has been delays caused by opposition from well-organized "intervenors," groups that take advantage of hearings and legal strategies to delay construction. Construction of the Midland, (Michigan) plant, referred to earlier in this chapter, was stopped for three years at one point, and delayed several times thereafter, which resulted in doubling the cost of the plant.[1] In the earlier Palisades plant built by the same uitlity, a $29 million addition was included in the $186 million plant to satisfy invervenors rather than incur further delay.

The Shoreham plant on Long Island was delayed[8] for three years by intervenors who turned the hearings for a construction permit into a circus. It included a total imposter claiming to be an expert with a Ph.D. and an

M.D., endless days of reading aloud from newspaper and magazine articles, interminable "cross examination" with no relevance to the issuance of a construction permit, and an imaginative variety of other devices to delay the proceedings and attract media attention.

The Seabrook Plant in New Hampshire suffered two years of delay due to intervenor activity.[9] The cost of this delay was estimated in 1978 as $419 million; with the unexpectedly high inflation rates since that time, this cost has undoubtedly risen substantially. It seems fair to classify additional costs caused by delay due to intervenors as regulatory turbulence as this is accomplished under regulatory procedures.

A rather different source of cost escalation is cash flow problems for utilities. When they institute a project, utilities do financial as well as technical planning. If the financial requirements greatly exceed what had been planned for, they often have difficulty in raising the large sums of extra money needed to maintain construction schedules. They therefore slow down construction or temporarily discontinue it, which greatly escalates the final costs of the plant. This is to a large extent due to regulatory turbulence which unexpectedly drove the costs up so high.

A final source of cost escalation is mistakes on the part of the constructors. This is what the antinuclear activists would have us believe is the principal cause. However, they are the same constructors that built nuclear power plants so efficiently and successfully in the 1960s and early 1970s. Are we to believe that after gaining all that experience they have become incompetent? I have no doubt that mistakes are made—they always have been and always will be in construction (or in any other human endeavor). If the frequency of mistakes has escalated, I strongly suspect that this is largely due to the constant design alterations engendered by regulatory turbulence.

In summary, there is a long list of reasons why the costs of nuclear plants have been higher than those estimated at the time the project was initiated, as shown in Fig. 20. Nearly all of these reasons other than unexpectedly high inflation rates, are closely linked to regulatory ratcheting and the turbulence it has created.

ACTUAL COSTS

The actual costs of nuclear power plants[10] in the United States (corrected for size to one million kilowatts) is plotted in Fig. 23 versus time of completion. Also shown there as crosses are the expectations of the United En-

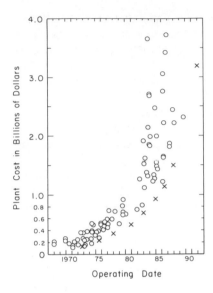

Fig. 23. Actual costs of nuclear power plants (per million kilowatts of output) versus their date of commercial operation (circles), from Ref. 10. Crosses show cost estimates versus predicted completion dates from United Engineers data used in Figs. 21–23.

gineers at the time of project initiation[3]; for example, the cross for 1978 is based on a United Engineers study in 1971 for a plant started in 1971 which they estimated would be completed in seven years. The principal reasons why the actual plant costs have been higher than those estimates are those outlined in the previous section, plus the fact that the inflation rate has been higher than was estimated at the time of project initiation.

One striking aspect of Fig. 23 is the wide variability of costs from one plant to another. Almost every nuclear power plant built in the United States has been custom designed. This is partly due to the fact that nuclear power is a young and vibrant industry in which technical improvements are frequently made. It is also due, to some extent, to regulatory ratcheting which causes the design to change continuously.

The variations in cost from one plant to another have many explanations in addition to difference in design. Labor costs and labor productivity vary from one part of the country to another. Some constructors adjusted better than others to regulatory ratcheting[11]; some maintained very close contact with NRC and were able to anticipate new regulations, while others tended to wait for public announcements. Many different designs are used for containment buildings, and reactors that happened to have small ones had much more difficulty in fitting in equipment required by new regulations. Some plants have been delayed by intervenors while others have not. Some have had construction delays due to cash flow problems of utilities. Plants nearing completion at the time of the Three Mile Island accident were delayed up to two years while NRC was absorbing the lessons learned from that accident and deciding how to react to them. These are some of the factors that have caused the wide cost variations in Fig. 23.

Not only have the costs been much higher than the United Engineers estimates, but construction times have been much longer—indeed that is part of the reason for the higher costs. The four U.S. nuclear power plants completed in 1981 were originally scheduled for completion[12] in 1973–1976; they were started[13] in the late 1960s when United Engineers was estimating completion times of 5–6 years. In short, experience with recently completed plants shown in Fig. 23 has been a disaster from the cost viewpoint, and things can be expected to be worse for plants completed over the next few years.

Figure 23 shows that between 1971 and 1983, the average cost of plants being completed increased by a factor of 10. During this period, the Consumer Price Index increased by a factor of 2.5, and the 2% per year escalation of construction costs over the CPI raises this to a factor of 3.2 that can be blamed on inflation. The explanation for the remaining factor of $(10 \div 3.2 =)3.1$ is in the direct and indirect costs of regulatory ratcheting.

With the cost situation as it now stands, no utility can afford to order a new nuclear power plant. Even if they accepted the United Engineers estimate in Fig. 20, the cost would be higher than most utilities can afford; the record of underestimates by United Engineers and the obvious explanations for it outlined in the last section give reason to fear that it might be much higher. Moreover, the long time required to build a plant—12 years according to United Engineers estimates (Fig. 22) and the past history of these estimates indicating that it may well be much longer—is too long for any business enterprise to do reasonable planning. It is obvious why no new nuclear power plants have been ordered since 1978 and why earlier orders have been canceled in droves. The antinuclear activists have won their battle, stopping the growth of nuclear power dead in its tracks. Plants now in the advanced stages of construction will be completed, but no new plants will be started in this country unless there are drastic changes.

On the other hand, if the worst excesses of regulatory ratcheting could be undone, reducing the construction time to 6–8 years and cutting the cost in half, nuclear power would be extremely attractive, by far the cheapest technology for generating electricity. The villain of our story is this regulatory ratcheting and the regulatory turbulence that has resulted from it.

THE SITUATION IN OTHER COUNTRIES

While the expansion of nuclear power has ground to a halt in the United States since the mid-1970s because of skyrocketing costs and lengthened construction periods, the situation is different in other parts of the world. The Soviet Union has commissioned their giant Atommash fabrication plant to turn out eight complete power reactor systems each year.[14] France started construction of seven reactors in 1979, one in 1980, and six in 1981, with plans to have them in operation in six years.[13] The French put 12 nuclear plants into operation during 1981 after an average construction time (following ground-breaking) of six years, whereas United States put a total of 6 plants into operation in 1980 and 1981 combined with an average construction period of 10.5 years. The French plants started in 1974–1975 cost an average of $500 million (per million kilowatts)[11]; none of the U.S. plants started at that time is yet completed, but when finished they will cost about $1500 million ($1100 million in 1981 dollars). An official French government estimate is that plants being started in the near future will cost $700 million,[15] whereas equivalent plants in the United States will cost $1700 million,[16] with both costs in 1981 dollars.

French plants, it should be noted, are built according to American designs under license from Westinghouse, and have often used American construction supervisors. I have never heard a claim that French plants are less safe than U.S. plants. The French have also been active in the reactor export business; they sold four reactors to foreign countries[14] in 1981. American manufacturers, who had 70% of the international market in the early 1970s, sold none. In 1981, 38% of French electricity was generated by nuclear power (versus 13% in the United States) and by 1985 it will be nearly 60%.[17]

Japan has been a late-comer to the nuclear power enterprise but it has progressed rapidly. The six plants completed there since 1979 averaged less than seven years in construction.[13] Germany has faced substantial opposition from antinuclear activists, but three new plants were started in 1982.

There are antinuclear activists in every country, but in no other major nation have they been as successful as in the United States. The French, reacting from a long history of turmoil, strengthened their government to the point where it can withstand the pressure from activist groups, and the Russians never had such pressure to contend with. It looks like those nations will end up having the cheapest electricity in the world while the United States will have the most expensive among large industrialized nations. This reverses the long-standing situation in which the United States had the cheapest power. Many economists believe that that has had a great deal to do with the past economic success of this country. If they are right, we are in for some hard economic times starting in the 1990s. Twenty years ago we laughed at Kruschev's threat "we will bury you," referring to his predicted economic ascendancy of Russia over the United States. He may yet have the last laugh.

THE POLITICAL BATTLE LOST

If regulatory ratcheting is to blame for halting the expanded use of nuclear power in the United States, it is only natural to ask who or what is to blame for regulatory ratcheting? We have already demonstrated that it has *not* been driven by new technical or scientific information; on the contrary the net effects of new information have been favorable and would suggest a *relaxation* of regulations. The source of regulatory ratcheting is clearly in the political arena. It is the result of nuclear scientists (including engineers and other technologists) suffering a crushing political defeat at the hands of the antinuclear activists. Let me give my own personal version of how this has come about.

First let's consider the cast of characters in the battle. The two sides are of an entirely different ilk. One of the main interests in life for a typical antinuclear activist is political battling, while the vast majority of nuclear scientists have no inclination or interests in political battling, and even if they did they have little native ability or educational preparation for it. While the typical antinuclear activist was taking college courses in writing, debate, and social psychology, the typical nuclear scientist was taking courses in calculus, radiation physics, and molecular biology. After graduation, the former gained wordly experience by participation in political campaigns, environmental activism, and anti-Vietnam War protests (I do not infer that these are not worthy causes), while the latter was gaining scientific experience working out the mathematical complexities in theories of neutron transport or radiation carcinogenesis, or devising solutions to some of the multitudinous technical problems in power plant design. While the former was making political contacts and developing know-how in securing media cooperation, the latter was absorbed in laboratory or field problems with no thought of politics or media involvement. At this juncture the former went out looking for a new battle to fight and decided to attack the latter; it was like a lion attacking a lamb.

Nuclear scientists had long agonized over such questions as what safety measures were needed in power plants, and what health impacts their radioactivity releases might cause. All the arguments were published for anyone to see. It took little effort for the antinuclear activists to collect, organize selectively, and distort this information into ammunition for their battle. Anyone experienced in debate and political battles is well prepared to do that. When they charged into the battle wildly firing this ammunition, the nuclear scientists first laughed at the naivety of the charges, but they didn't laugh for long. They could easily explain the invalidity of the attacks by scientific and technical arguments, but no one would listen to them. The phony charges of the attackers dressed up with their considerable skills in presentation sounded much better to the media and others with no scientific knowledge or experience. When people wanted to hear from scientists, the attackers supplied their own—there are always a few available to present any point of view, and who was to know that they represented only a very tiny minority of the scientific community. The antinuclear activists never even let it be made clear who they were and whom they were attacking. The battle was *not* billed as a bunch of scientifically illiterate political activists attacking the community of nuclear scientists, which is the true situation. It was rather represented as "environmentalists"—what a good, sweet, and pure connotation that name carries—attacking big business interests (the nuclear industry) which were trying to make money at the expense of the public's health and safety. Jane Fonda refused to debate nuclear scientists; her antagonists, she said, were corporation executives.

The rout was rapid and complete. In fact the nuclear scientists were never even allowed on the battlefield. The battlefield here was the media which alone have the power to influence public opinion. The media establishment swallowed the attackers' story, hook, line, and sinker, becoming their allies. They freely and continually gave exposure to the antinuclear activists but never gave the nuclear scientists a chance. With constant exposure to this one-sided propaganda, the public was slowly but surely won over. Some of the details have been covered in earlier chapters—how the public was driven insane over fear of radiation (cf. Chapter 2); how it became convinced of the utterly and demonstrably false notion that nuclear power was more likely to kill them than such well-known killers as motor vehicle accidents, cigarette smoking, and alcohol (cf. Chapters 1 and 4); that burying nuclear waste, actually a very simple operation, was one of the world's great unsolved problems (cf. Chapters 5 and 6); that contrary to all informed sources, the Three Mile Island accident was a close call to a disaster (cf. Chapter 3); and so on. Fears of everything connected with nuclear power were blown up completely out of perspective with other risks (cf. Chapter 4). Hitler's man, Goebbels, had shown what propaganda can do, but nuclear scientists never believed that it could succeed against the rationalism of science; yet succeed it did. The victory of the antinuclear activists was complete. Even conservative businessmen and staid housewives have been converted and support them. A few nuclear scientists, like myself, have been trying to fight back—this book represents my dedication to continuing the fight—but without access to the media, there is little we can do. For the past eight years, I've been giving about 40 public lectures a year, talking to any group that invites me. I might address a total of 4,000 people a year with all this exhausting effort, but one TV program communicates with 40 million, ten thousand times as many. They are professionals at gripping an audience, they have large budgets, personnel, and tremendous facilities. What chance does scientific evidence have against such forces?

Having won the battle for the minds of the American public, the antinuclear activists are in control. When Ralph Nader writes a letter, the Nuclear Regulatory Commission jumps. But political muscle is not even necessary. With the media and the public banging on the door, NRC has had no choice but to ratchet its regulations. The antinuclear activists have won their battle, and to the victor belong the spoils—the failure of nuclear science to provide the cheap and abundant energy our nation so sorely needs. That is the goal they cherished and they have achieved it. Our children and grandchildren will be the victims of their heartless tactics. When Shakespeare said "the truth will out," he didn't reckon with the power of the modern media.

COST PER KILOWATT HOUR

The cost of electricity produced by a power plant depends on many things besides the initial cost of the plant. These are conveniently grouped into three categories:

(1) Fixed charges, which cover items that do not depend on the quantity of electricity produced, like amortizing the construction cost of the plant, major maintenance problems that must be undertaken once or twice during the plant's lifetime, insurance, taxes, dividends for utility stockholders, and ultimate decommissioning of the plant. The fixed charge is determined by assuming that it is used to purchase an annuity which will pay off all of these costs, including interest on them, by the end of the plant's useful life. (For expenses occurring near the end of the plant's life, like decommissioning, it is assumed that interest is accrued from the annuity.)

(2) Operating and maintenance, which is primarily the salaries and wages paid to plant personnel, but includes materials, supplies, and outside services.

(3) Fuel costs, including costs for purchase of the fuel and management of the spent fuel and the waste it generates.

The sum of these three charges for one year is then divided by the number of kilowatt-hours (kW-hr) of electrical energy the plant produces during the year, to obtain the charge to the customer per kW-hr (this is the cost/kW-hr at the plant; it does not include the distribution cost).

According to the mid-1982 analysis[16] by the U.S. Energy Information Administration (EIA), a nuclear plant started at that time and completed in 1995 would price its electricity at

Fixed charges:	2.9¢/kW-hr
Operation and Maintenance:	0.5¢/kW-hr
Fuel:	0.85¢/kW-hr
Total:	4.3¢/kW-hr

in which all costs are expressed in 1980 dollars. An estimate by a construction firm economist[5] gives a similar result.*

To give perspective on this 4.3¢/kW-hr cost, it is useful to trace the history of the cost/kW-hr of electricity in the United States. It declined steadily from the 1880s when electricity first came into use until 1970 when it reached a low of 1.9¢/kW-hr (in 1980 dollars). By 1976–1977 it rose to 2.1¢/kW-hr for nuclear and 2.4¢/kW-hr for coal, and by 1980 it had climbed to 2.3¢ and 2.5¢, respectively. (Recently completed French nuclear plants are now producing electricity[18] for 1.7¢/kW-hr in 1980 dollars.) The EIA estimate above therefore indicates that the cost of nuclear electricity from a plant started now and completed in 1995 would be nearly double the present cost. A study by the European Economic Community[19] gives as typical a plant started in 1982 and completed in 1990 (5 years earlier than the American plant started at the same time), producing power at costs in 1980 dollars of 2.3¢/kW-hr in France, 2.2¢/kW-hr in Italy, 2.8¢/kW-hr in Belgium, and 3.1¢/kW-hr in Germany. Here again, we see a foreboding reversal of the historic situation where electricity has been consistently cheapest in the United States.

From the breakdown given above we see that 2/3 of the cost of electricity is due to the fixed charge, which is largely based on the cost to construct the plant. We have already assigned the blame for that high cost to regulatory ratcheting. But regulatory ratcheting influences the cost of electricity in other ways. By requiring frequent shutdowns for safety-related purposes such as inspections or modifications, it reduces the number of kilowatt-hours gen-

---

* Actually Ref. 5 gives 6.7¢/kW-hr as the charge during the early years of operation of the plant. This is due to the normal practice of rapid recovery of investment followed in the United States whereby the real cost/kW-hr from a plant decreases with time. The EIA analysis gives the average real charge over the life of the plant.

erated per year by a plant. This increases the cost/kW-hr proportionately. One consequence of regulatory ratcheting has been to increase the work force required at plants, which reflects directly in the operating-maintenance costs. Here again we find that regulatory ratcheting has been the villain, driving up the cost of nuclear electricity, making it much more expensive in the United States, the land where it was pioneered and developed, than anywhere else in the industrialized world.

COAL VERSUS NUCLEAR COSTS

One possible answer to keeping U.S. electricity cost competitive might be building coal-fired plants instead. We have shown that this is harmful from the health standpoint—every time a coal-burning power plant is built instead of a nuclear plant (which has been happening about ten times each year), several hundred extra people are condemned to a premature death. But the antinuclear activists don't worry about the few thousand premature deaths per year they are causing. Maybe that's not too big a price to pay to keep our nation economically competitive if nuclear power is rejected.

Unfortunately, however, the cost of coal-fired electricity has been escalating almost as fast as that of nuclear electricity. Construction costs have risen steeply as more pollution abatement equipment has been added, and operation and maintenance of this added equipment are also expensive. Fuel costs, i.e., the cost of coal, are especially subject to inflation. EIA estimates[16] that the price of coal will increase at an average rate of 3.3% per year faster than the general inflation rate (CPI). Their breakdown of costs in 1980 dollars for electricity from a coal-burning plant completed in 1995 are:

Fixed charges:	1.7¢/kWhr
Operation and Maintenance:	0.4¢/kWhr
Fuel:	2.2¢/kW-hr
Total:	4.3¢/kWhr

We see that this is just equal to their estimate of the cost of nuclear electricity. The analysis by the construction firm economist[5] found the cost/kW-hr for coal to be 13% higher than for nuclear energy. Coal is therefore not the solution to our economic problems with the cost of electricity.

Many analyses have been made of cost comparisons between nuclear and coal plants with construction started at the same time. This is clearly a very practical question for a utility that decides it must build a new power plant. There are analyses by economics consulting firms they hire, from banking organizations which maintain expertise to aid in decisions on investments, and from utilities themselves. Since the early 1970s, they all consistently found that nuclear power is the cheaper of the two. For example, the Tennessee Valley Authority (TVA) is the largest electric utility in the United States. Its profits, if any, are turned back to the U.S. Treasury. It maintained a large and active effort for many years in analyzing the relative cost advantages of nuclear power versus coal, consistently finding that nuclear power was cheaper. The 1982 analysis by the Energy Information Administration was its first to find that coal and nuclear power are equal in cost; their previous analyses found nuclear power to be cheaper.[4]

The only exception of which I am aware is Charles Komanoff,[20] who consults not for utilities but for antinuclear activist organizations. He has consistently claimed that, contrary to past experience, coal-fired electric power will be cheaper than nuclear power for plants started at the time of his analysis. His conclusions have been based on some rather unusual assumptions.[21] He has always assumed that the capacity factors, the ratio of electricity actually produced to what would be produced if the plants operated full time at maximum capacity, would be higher for coal than for nuclear power, whereas in past experience, the opposite has been the case. In his 1976 analysis, when all other analyses were finding a large cost advantage for nuclear, he assumed a capacity factor of 48% for nuclear power, although that figure was exceeded by 36 of 38 nuclear plants in his data base, while he used a 73% capacity factor for coal, which was higher than that experienced by 86 of the 93 plants in his data base. In his 1980 analysis, when other analyses were finding nuclear power's cost advantage to be much smaller, he used 60% for nuclear power and 70% for coal. One cannot help but suspect some lack of objectivity in his work.

Komanoff assumes that regulatory ratcheting will continue unabated for nuclear plants, with no comparable effects on coal plants. He assumes that the price of a ton of coal will not rise appreciably in the future. He uses many analytical procedures that professionals long involved in the business object to strenuously.

Since I am not a professional economist, and I have never carefully studied the details of an economic analysis, my view of this question must be largely based on a decision of whom to believe. The fact that Komanoff's views are contrary to those of the great majority of economics professionals impresses me. The fact that the latter are involved in actual decision making on how to spend billions of dollars, while Komanoff's analyses are used only for political purposes impresses me much more.

There have been claims that utilities favor nuclear because they can derive more profit from it. But this would not explain the findings of non-profit utilities like TVA, or the municipally owned utilities. It would certainly not explain why the Soviet Union and other Communist bloc nations are turning to nuclear power, while they have the world's largest coal resources. Most Western European power plants are built and owned by their governments.

Further support for the economic benefits of nuclear power relative to coal comes from a study by the European Economic Community.[18] For plants completed in 1990, it finds that nuclear power costs will be cheaper than those from coal-fired plants by a factor of 1.7 in France, 1.7 in Italy, 1.3 in Belgium, and 1.4 in Germany. These results are based on only a very slow rise in the price of coal, and do not include scrubbers for removal of sulfur. Scrubbers are required on new coal-burning plants in the United States and add very substantially to their cost.

All of this convinces me that Komanoff is wrong, and that nuclear power has had a cost advantage over coal, although by now it may have largely been dissipated in the U.S. However, the media rarely present that viewpoint. Komanoff has received more media coverage than all of his critics and all of the other involved economics analysts combined. What happened with radiation scientists, where the tiny minority who support the antinuclear activists' viewpoint got all the media coverage, is apparently happening again with economists. Komanoff has become a king pin of the antinuclear movement, frequently speaking on their public information programs. Questioners from my audiences express complete surprise when I point out, at least until recently, the great majority of economics analysts have consistently found nuclear power to be cheaper than coal—all they had ever heard was the Komanoff viewpoint.

This leads us to the question: why have utilities been ordering coal plants rather than nuclear plants for the past 5–8 years? The economic difference between the two is that coal burning plants are cheaper to build, but they have much higher fuel costs, which raises the total cost of the electricity they produce to above that of nuclear plants. In present circumstances, utilities have great difficulty in raising capital to build new plants, so it is easier for them to opt for coal burners. Of course they will later have to spend a lot more money on fuel, but that will be paid for directly by their customers at the time it is spent. The utility suffers less financial strain that way, with their customers later paying more for their electricity.

The easiest path for a utility these days is not to build *any* new power plants which, to a large extent, is the policy they have been following. However, our population is expanding and our economy must somehow develop more than a million new jobs each year for the rest of this century as today's children grow to adulthood. It therefore seems reasonable to expect growth in our electric power needs. This requirement is not so evident as long as our economy is depressed, but if rapid economic growth and prosperity do return, they may be seriously hampered by a lack of electric power.

If that situation does develop, there will be little that can be done about it because many years are required to plan and construct a new power plant. Many economic planners therefore feel that the current stagnation in new power plant construction is dangerous. They point out that the economic penalty in building more plants than needed, is much less than in needing more electricity but not having it available. But in the former case, the utility suffers much of the loss while in the latter it is imposed on the entire nation.

It is not fair to blame the utilities for not building more plants. They have often proposed to rectify this situation by charging their customers for construction work in progress, but public utility commissions have rarely allowed them to do so. We therefore now enjoy cheap electricity from plants built many years ago. When these plants wear out, or if our expanding economy requires more power, our electric bills will increase dramatically. Or worse still, the needed electricity may simply not be available, and our children will have to live with high unemployment and poverty.

Over and above these considerations, it would be desirable to replace oil- and gas-burning power plants to preserve those scarce resources. As they now generate about 30% of our electricity, replacing them with nuclear or coal-burning plants would save a great deal of oil and gas. Since we have enough oil and gas for now, however, we are doing little about this problem. Again, we are not suffering as a result of the political conquests of the antinuclear activists, but our children will pay the price in redoubled terms.

It is not too late to rectify the situation. We could reverse the regulatory ratcheting and allow nuclear power plants to be built rapidly, efficiently, and cheaply—as well as safely. The French technology was copied from our own, on a licensing agreement with Westinghouse, so there can be no question but that we can do what they are doing. We could cut our future cost of electricity in half. What a boon this would be to our industry as well as to our household budgets! But to accomplish this, the public must be cured of its insane fear of radiation, and the media must stop transmitting the false message that nuclear power is dangerous. The public misunderstandings outlined in Chapters 2–6 must be corrected. It sounds like an impossible task, but with so much at stake, we must try as hard as we can.

REFERENCE NOTES

1. Consumers Power Co., Jackson, Michigan, Position Papers (1981); R. Wheeler, "The Midland Chronicles," Consumers Power Co., Jackson, Michigan (1981).
2. W. B. Behnke, "Economic and Technical Experience of Nuclear Power Production in the U.S.," International Conference on Nuclear Power Experience, Vienna (1982).
3. United Engineers and Constructors, "Final Report and Initial Update of the Energy Economic Data Base Program" (1979); also Arthur G. Bernstein (United Engineers), private communications (1982).
4. A. Reynolds, "Cost of Coal vs. Nuclear in Electric Power Generation," U.S. Energy Information Administration Document (1982).
5. W. W. Brandfon, "The Economics of Nuclear Power," American Ceramic Society, Cincinnati (1982).
6. I. Spiewak and D. F. Cope, "Overview Paper on Nuclear Power," Oak Ridge National Laboratory Report ORNL/TM-7425 (1980).
7. J. H. Crowley, "Nuclear Energy—What's Next," Atomic-Industrial Forum Workshop on The Electric Imperative, Monterey, California (1981).
8. Long Island Lighting Co., "The Shoreham Nuclear Power Plant: An Overview" (1982).
9. M. R. Copulos, "Confrontation at Seabrook," The Heritage Foundation (1978).
10. Tennessee Valley Authority, "Summary of Capital Cost per KW for U.S. Nuclear Plants" (1982).
11. S. Diamond, "Shoreham: What Went Wrong," Newsday, November 18 (1981).
12. "The World List of Nuclear Power Plants," Nuclear News, February (1982), p. 83.
13. Nuclear Engineering International, August (1982), Supplement.
14. Atomic Industrial Forum, "Nuclear Power: Facts and Figures," July (1982).
15. Journal Official des Debats de l'Assemblee National, May 10 (1982).
16. "Projected Costs of Electricity from Nuclear and Coal-fired Power Plants," U.S. Energy Information Administration Document DOE.EIA-0356/1, August (1982).
17. OECD Nuclear Energy Agency, Tenth Annual Activity Report (1982).
18. Electricite de France/Production Thermique—Annual Report (1982).
19. International Union of Producers and Distributors of Electrical Energy Conference in Brussels, June 1982. Reported in Nuclear News, July (1982), p. 48.
20. C. Komanoff, "Power Plant Cost Escalation: Nuclear and Coal Capital Costs, Regulation, and Economics," Komanoff Energy Assoc., New York (1981).
21. A. D. Rossin, "Statistics, Economics, and Credibility," Nuclear News, September (1981); AIF Industry Review Group, "Critique of the Charles Komanoff Report," Atomic-Industrial Forum, Washington, D.C. (1981).

Chapter 9 / THE SOLAR DREAM

The last of our major points of public misunderstanding is the widespread impression that solar electricity will soon be available, so there is no reason to get involved with nuclear power. There are vociferous political organizations pushing this viewpoint, and the media give it powerful support. The public seems to be largely convinced that our primary source of electricity in the next century will be solar.

As a result of my frequent participation in scientific meetings on energy technology, I have come to know several solar energy experts, but I have yet to meet one who shares that opinion. Their professional lives are devoted to development of solar electricity, and most of them are very enthusiastic about its future. Nevertheless they have frequently encouraged me in my efforts on behalf of nuclear power, saying frankly that the public's expectations for solar power are highly unrealistic. They foresee its future as a *supplement* to other technologies, with advantages in certain situations, rather than as the principal power source for an industrialized society.

There are two major obstacles that prevent solar electricity from playing the latter role: it will be very expensive by present cost standards, and its availability is highly irregular. Our discussion of these problems will be complicated by the fact that several different types of solar electricity are under development, but we will cover what are popularly regarded as the three most important of them: (1) photovoltaics, popularly known as solar cells; (2) the power tower, in which heat from the Sun is used to boil water into steam which is then used to generate electricity as in a coal-fired or nuclear power plant; and (3) wind turbines, an elaboration of the familiar windmill which has provided some of our power for many centuries.* Note that we are not considering applications of solar energy to heating buildings or providing hot water since they are essentially irrelevant to the topic of this book. Nuclear power is most useful for generating electricity while only a small fraction of all electricity is used for those purposes, generally in situations where convenience is considered more important than costs.

COST PROBLEMS

The United States Department of Energy has a large program in photovoltaic development, including an ongoing analysis of costs and realistic goals for future costs. With the technology now in hand, it is estimated[1] that solar cells can be produced for $2.80 per peak watt and that a complete solar electricity system would cost $10 per peak watt. A peak watt is the electric power output in watts at noon on a sunny day; all costs are in 1980 dollars. The goals of the program are to develop the capability of producing solar cells for 70¢ per peak watt by 1987 and 25¢ per peak watt by the end of the development program in the 1990s, and of making complete systems for $1.60 per peak watt by 1987 and $1.20 per peak watt in the 1990s.[1,2]

The difference between the cost of the cells and the complete system consists of a number of items, referred to as "balance of plant." These include[3,4] engineering, site preparation, structural supports (typically 30¢ per peak watt[5]), installation, electrical hook-up of cells (8¢), conversion to alternating current with proper voltage and frequency (25¢), synchronization with interconnected power sources (8¢), and management, sales, profits, and other administrative

---

* Since winds are driven basically by forces generated by the Sun's heat, wind power is often classified as solar energy. Many other potential technologies for generating electricity could be discussed here—ocean thermal gradient, satellite solar, geothermal, tidal, ocean waves, biomass, hydrogen, and synthetic fuels to name the most prominent examples. However, none of these is generally considered to be more promising than the three technologies discussed here, and to include them all would be a very substantial diversion from the subject matter of this book.

charges. These add up to about 90¢–$1.00 per peak watt. Fairly intense efforts over several years have not substantially reduced them.[6]

Note that no costs have been included for interest on funds used during construction, which was a very expensive item contributing to the costs of nuclear plants. The rationale for ignoring it here is the supposition that solar photovoltaic plants can be built and put into operation in the very short time of one year or less.

The amount of sunlight falling on the cells varies throughout the day (becoming zero at night), changes markedly between summer and winter, and is sometimes reduced by cloud cover, so the average output is many times less than the peak, about 4.5 times less in the Southwest desert, and 7 times less in the Northeast. The cost per *average* watt of available power production, if the goals of these programs are attained, will therefore be something like $7.

Let's try to understand these costs a little better. The amount of sunlight falling on the earth's surface at noon on a sunny day is about 1000 watts per square meter, or 100 watts per square foot. Photovoltaic systems can never be much more than 10% efficient in converting solar energy to electricity— including losses in covers, frames, and electrical conversion, present systems are only 5%–8% efficient.[7] We can therefore expect to obtain a peak electrical output of no more than 10 watts per square foot. The goal of our national photovoltaic development program on which the prices quoted above are based is to produce solar cells for 25¢ per peak watt, which means ($0.25 × 10 =) $2.50 per square foot.

An ordinary cement sidewalk costs $2 per square foot[8]; this means that our price estimates for solar electricity are based on covering the ground with solar cells for practically the same price as we pay to cover it with a thin layer of cement. A solar cell is a highly sophisticated electronic device, about one inch in diameter, based on advanced principles of quantum physics developed in the 1950s. It is made from materials of extremely high purity, a purity that was unattainable even in scientific laboratories until the late 1940s. Cells are manufactured by a process that has taken some of the best efforts of modern technology to develop. They must be capable of standing up to all the vagaries of outside weather for 20–30 years. To cover the ground with these sophisticated devices for roughly the same price we pay to cover it with a thin layer of cement, manufactured simply by grinding up rock and heating it, is indeed a challenging goal. Surely we cannot hope to do much better.

The numbers quoted here also explain why solar electricity is so expensive—the energy in sunlight is spread so diffusely that we must collect it from large areas with correspondingly large collectors in order to obtain appreciable amounts of power. To produce the power generated by a large nuclear plant would require covering an area 5 miles in diameter with solar cells.

This discussion has been in terms of a large central electricity generating facility; at least equal consideration should be given to individual rooftop installations on houses. There is every reason to believe, however, that these will be equally or more expensive.[9] Costs for land and structural support may be reduced, but there would be increased costs for installation, maintenance, sales commissions, distribution, power conditioning, safety features, controls and displays, and insurance. In addition, there are complications that might add to the expense from building codes, product standards and liabilities, solar access, zoning restrictions, aesthetics, utility system maintenance, operation, emergency procedures, utility interconnection, and installer training. Retrofitting existing houses is substantially more expensive than installation during initial construction. For apartment buildings and most commercial and industrial buildings, there is not sufficient roof space to provide the needed electricity. The $7 per average watt cost for a central installation is therefore as low a cost as can be realistically considered as a national average.

According to the U.S. Energy Information Agency,[14] the cost of a nuclear plant started in 1982 and completed in 1995 is $1.60 per watt in 1980 dollars (one-billionth of the cost of the one billion watt plants discussed in Chapter 8). Since it is available for operation about 72% of the time,[11] this represents a cost of ($1.60 ÷ 0.72 =) $2.20 per average available watt. Nuclear plants require fuel which contributes about 20% to the cost of their electricity; to compensate for this difference we should increase the equivalent cost of a nuclear plant by 20%, to $2.75. If we assume that operation and maintenance costs are similar for nuclear and solar plants (in any case they represent only a small part of the total), this $2.75/average watt for nuclear power is directly comparable to the $7/average watt for solar electricity.

Even this argument understates the problem with solar costs because the above cost of nuclear plants is already too high for utilities to accept; they are not ordering them. It would probably be more meaningful to compare them with the equivalent costs of French plants[12] which will be about $1.35 per average watt.* The costs in 1980 dollars of U.S. nuclear plants completed in the 1975 time period, which determine the price we are now paying for electricity, were about $1.10 per average watt* (cf. Chapter 8, Fig. 4). I have

---

* Numbers given here are cost of plant ÷ 0.72 for availability and × 1.4 to represent the cost of fuel.

rarely seen estimates of future costs per kilowatt-hour for photovoltaic electricity. One that has appeared in print is 8.5¢/kW-hr in 1980 dollars.[13] The average cost for coal and nuclear generated electricity in 1980 was 2.4¢/kW-hr at the generating plant. It is difficult to understand the fact that these cost/kW-hr differ only by a factor of 3.5 whereas the cost of the plant plus fuel differs by a factor of six. I cannot help but suspect some accounting hocus-pocus. In any case, it is clear that solar photovoltaic electricity will be many times more expensive than the price we are accustomed to paying, or the price Europeans will be paying for nuclear electricity in the 1990s, 2.2¢–3.1¢/kW-hr.[14]

The U.S. Department of Energy also has a large project for developing solar electricity with the power tower approach.[6,15] It is now completing a 75 acre pilot plant near Barstow, California with nearly a million square feet of glass mirrors tilting under computer control to reflect sunlight to a receiver at the top of a 300-foot-tall tower. A plant this size could provide a peak power about 1% of that from a large nuclear or coal-burning plant. The Barstow plant costs $11 per peak watt, but the long-range goal of the Department of Energy program is to reduce the cost of power tower plants down to about $1.80 per peak watt.[6] These plants produce no power when the sun is covered by clouds, so this corresponds to no less than $8 per average watt even in the most favorable locations. This is slightly higher than the hoped-for future cost of photovoltaic systems.

Costs are considerably more favorable for wind turbines,[18] at least in some locations. Where the *average* wind velocity is at least 15 miles per hour, there is reason to believe that wind can be made cost competitive, per average watt, with coal-fired or nuclear electricity. This gives wind a very large economic advantage over other forms of solar electricity, and indeed there is active planning for commercial ventures in large-scale windpower in some locations (small-scale applications have been in operation for many years).[19]

One disturbing aspect of this technology is its profligate use of land—in a favorable location a 200 square mile area containing a thousand towers at least 200 feet high is needed to produce the same electrical output as a single large nuclear plant (which occupies less than one square mile). In most places, the wind velocity is much lower, making this technology far more expensive and land consumptive.

But the greatest difficulty in depending on wind, or on solar cells or power towers, as our principal source of electricity is in the variation of their availability.

Is It There When We Need It?

Since electrical energy is difficult to store, it must ordinarily be used as it is produced. Our principal uses of electricity, unfortunately, do not vary in time the same way as sunshine and wind vary. There is little seasonal variation in our use of electricity, but the influx of solar energy is 2 to 3 times higher in summer than in winter.[20] We use a great deal of electricity at night when there is no sunshine, and there is usually little reduction in our electrical needs on cloudy days. If we were largely dependent on solar energy, how would we handle these problems?

The most obvious solution is *storage*, and in this connection we first think of storage batteries like those used in automobiles. These could solve some of the short-term problems; for example, $3,000 worth of batteries replaced every 2–3 years could be charged during the day enough to handle ordinary nighttime uses in a home without air conditioning.[21] But several cloudy days in a row would cause considerable hardship. The summer–winter variations are much more difficult. One could, of course, have a system with enough capacity to provide ample power with storage batteries for winter, even including cloudy periods, but its capacity would have to be five times that needed to provide the annual average power used[21]; thus it would cost five times as much as the systems we have been discussing. Electricity would then be about 20 times more expensive than it is now! The problem would be much more difficult for businesses and factories that use a lot of electric power at night and in winter.

The only other practical storage system for electricity is using it to pump water up to a reservoir on a hill; it can then generate electricity when it is allowed to flow back down.[22] This is expensive to construct, wastes about one-third of the electricity, has various environmental problems,* and is applicable only in areas with plenty of water and hills. It would not be practical for rooftop installations.

Aside from storing electricity, the other simple solution to variations in availability is to have back-up sources of electric power. One might consider having a nuclear or coal-burning power plant available, but this would make no sense economically. The standby plant would have to be constructed, and it would need nearly a full complement of operating and maintenance per-

---

* By far the most important project of this type ever undertaken, the Storm King reservoir on the Hudson River north of New York City, was delayed for many years and finally abandoned because of opposition from environmental organizations.

sonnel. The only saving by using solar energy would be in fuel costs, which represent only about 20% and 50%, respectively, of the total cost of nuclear and coal-fired power.[10] The cost of back-up power to the customers would have to be not much less than the cost of obtaining *all* their electricity from those plants.

We often hear stories about individuals with windmills or solar cells using a regular utility line for back-up power and even selling the excess power they generate at various times back to the utility—in many states utilities are required by law to purchase it. This does little harm as long as only a few individuals are involved, but it wouldn't work if a large fraction of customers did it. The utility would not only have to build and maintain back-up power plants without selling much of their product, but they would have to buy a lot of power they don't need when the sun is shining or the wind is blowing. The utility could only survive by raising the price of the little electricity it does sell sky high.

The most practical back-up power for a rooftop solar cell installation would be a diesel generator. This is a very expensive way to generate electricity, and if the price of oil rises as expected in the future it will become much more expensive. It also gives the home owner a lot of maintenance and repair work and its reliability leaves much to be desired. A diesel engine is also a very noisy device to have in the home.

In general, the amount of back-up or storage capacity needed to overcome the variable nature of sunshine and wind depends on how much inconvenience we are willing to endure. But even to approach the dependable electrical service we now enjoy would be extremely expensive.

WHY SOLAR ELECTRICITY?

With all these problems, why are many good and honest scientists pushing solar electricity, devoting their careers to its development? The answer is that it may fit in as a *supplementary* source. There are already some situations in which it is useful. In a remote area not serviced by a utility, electricity is often obtained from a diesel-engine-driven generator fueled by oil that must be trucked in. This electricity may cost as much as $1 per kW-hr, 20 times what an American urban-dweller pays, in which case solar electricity may be cheaper. Also, people in such situations may be willing to put up with considerable inconvenience from power outages. Another reasonable application is pumping water for irrigation, since the intermittent nature of solar energy often doesn't matter; when the sun shines, the pump runs, and when the sun doesn't shine, it stops, with no resulting harm or inconvenience.

If the price of solar electricity comes down to the hoped-for levels, many more applications will become practical. The most important of these is in providing "peak load" power for utilities.[7] The largest demand for electric power in most areas is on hot summer afternoons when lots of air conditioners are running. Utilities do not service this peak load with coal-fired or nuclear power plants because the latter, due to their high capital cost, are economically attractive only if they operate much of the time. Instead they use internal combusion engines which are cheap to purchase but expensive to run. This peak load electricity is thus much more expensive than that supplied steadily day in and day out, making solar power more competitive. Moreover, the intermittent availability of solar power is not so much of a problem here, as the air conditioning load is large mainly when the sun is shining.* This is widely regarded as the best opportunity for solar electricity to have important application.

Further opportunities depend on large increases in the future costs of coal and nuclear power. If the point is reached where solar electricity becomes cheaper than these, it is estimated[1] that its intermittent availability would not limit its usefulness unless it comprised more than 30% of our electricity. However, this still would require that most of our electricity be derived from coal and nuclear fuels.

There are some people who foresee the day when all our electricity will be solar, but they envision a very different world from our present one.[23] It is a world of low technology and a simpler life, a more desirable lifestyle, in their view. It will of necessity be a life of more unmechanized farming and manual labor, of fewer machines, comforts, and conveniences. They call it living in harmony with nature (but it might also be called sliding back toward the lifestyle of our primeval ancestors). In such a world they contend that there would be no place for large nuclear or coal-fired power plants, and little place for other large industrial operations except, presumably, for manufacturing solar cells.

The validity of their views depends on social, political, demographic, and psychological considerations on which I have no claim to expertise. I

---

* Actually, the maximum solar energy comes at noon while the maximum air conditioner use is at 2–3 PM and this difference does cause serious problems. People would have to precool by running air conditioners in the morning, and insulate to avoid running them in late afternoon and evening. This increases the amount of electricity needed.

only want it to be clearly understood that their ideas are driven by political considerations rather than by scientific, technical, or economic analyses. Their attempts at the latter have been shallow and heavily biased, and have generally received rebuttal rather than acceptance by experts.[24]

ENVIRONMENTAL PROBLEMS, THE MEDIA, AND POLITICS

In earlier chapters we have recounted how the media have spread misunderstanding about nuclear energy; their handling of solar energy has been equally biased, but in the opposite sense. I have never seen or read a media story that clearly points out the serious problems with solar electricity as explained in the first two sections of this chapter. Everything they present is positive, not to say propagandistic.

My favorite news story of this type was about a solar automobile with its roof covered by solar cells. The total solar energy falling on a car roof at noon on a sunny day in summer is about 1000 watts, and with the 10% maximum available system efficiency, this would give 100 watts of electricity which is about 1/8 horsepower. Even the smallest cars require 50 horsepower engines, so how can the solar car run with several hundred times less power? And what about early mornings, late afternoons, cloudy days, and winter, not to mention nighttime?

I have never seen or heard media coverage of health and environmental problems with solar energy. It was pointed out in Chapter 5 that production of the materials for deploying a solar cell array requires burning 3% as much coal as would be burned in generating the same amount of electricity in coal burning power plants. Roughly the same is true for the power tower and wind turbine applications of solar energy. That means that they produce 3% as much air pollution as coal burning, which is not a great environmental problem, but it still makes them more harmful to health than nuclear power. In addition there are long-term waste problems discussed in Chapter 5 which are hundreds of times more of a health problem than the widely publicized nuclear waste. There are lots of poisonous chemicals used in fabricating solar cells,[25,26] such as hydrofluoric acid, boron trifluoride, arsenic, cadmium, tellurium, and selenium compounds which can cause health problems in the event of an accident. Also there is much more construction work needed for solar installations than for nuclear; construction is one of the most dangerous industries from the standpoint of accidents to workers.

How many people would be killed and injured in cleaning or replacing solar panels on roofs, or in clearing them of snow? What about the dangers in repairing the complex electric conversion system? Over a thousand Americans now die each year from electrocution, and the power conditioning equipment needed for a solar electricity installation would represent a major increase in this risk. Back-up systems, most especially diesel engines in the home, have serious health problems.[27] Diesel exhausts include some of the most potent carcinogens known, and they have most of the other air pollution problems discussed in connection with coal burning in Chapter 4.

There are also environmental and ecological problems.[28] What happens to the land and animals that live on it when a 5-mile-diameter area is covered with solar cells or mirrors? Desert areas, which are most attractive for solar installations, are especially fragile in this regard.

Wind turbines are noisy and some consider them to be ugly, especially if they dominate the landscape for many miles in every direction—recall that it takes a thousand towers as high as a 20 story building covering 200 square miles to replace one nuclear plant. They further cause problems for birds and flying insects, and they interfere with television reception.[17,22] Power towers use a great deal of water, which is generally in short supply in deserts where these installations would be most practical. All solar electricity technologies use a lot of land, inhibiting its use for other purposes; in favorable areas a rooftop installation might satisfy the needs of a single-family house, but this is certainly not true for apartment buildings, office buildings, and most industrial plants.

These health and environmental problems are not terribly serious, but an objective analysis would rate them more harmful than the health and environmental impacts from nuclear power. However, the media constantly harp on the problems with nuclear power, never mentioning those of solar energy. Even minor technical difficulties in nuclear plants unrelated to safety frequently receive heavy media coverage. On the other hand, a great many solar cells have failed permanently within a few months after installation, and large wind mills have had catastrophic failures, but these are never mentioned in the media.

As a result of this imbalance in media coverage, the public considers solar energy to be the safest, soundest, and most environmentally benign way of obtaining electricity, and believes that nuclear power is on the opposite end of the spectrum. The public also seems to believe that solar electricity will be cheap and easy to use once it is developed. I know of no scientific basis for such optimism.

In practical terms, this favorable public image of solar energy has been very important politically. Our government is ever responsive to the public's desires. The federal government directly subsidizes 40% of the cost of any purchases of solar equipment as a reduction in tax payments. Several states offer additional subsidies leaving the customer to pay as little as 30¢ on a dollar.

Favoring solar energy was part of political life during the Carter Administration, while nuclear energy was officially labeled the "source of last resort." The Carter appointee as head of TVA, our nation's largest electrical utility, refers to the present time as "the presolar age." Government agencies stumbled over one another in rushing to distribute prosolar propaganda, while any public distribution of information favorable to nuclear power met with harsh criticism from Congressmen and administration officials. My use of the words "propaganda" and "information" here is intentional. I call the solar material "propaganda" because it creates false impressions by not mentioning the very serious cost and availability problems discussed above and by implying that our electricity has a largely solar future. The nuclear information, on the other hand, has generally been a balanced account of how the involved scientific community views the questions under discussion.

Congress established the Solar Energy Research Institute to lead the scientific and technological development of solar energy. The Carter administration appointment for the Director of this Institute of science and technology was a recent college graduate, a history major with no scientific or technical training or experience. His claim to fame was leadership in prosolar political groups. His first major action on assuming the Directorship was to fire some of the best scientists in the Institute because they were not sufficiently enthusiastic about solar energy. All of the other National Laboratories and Research Institutes in the United States are, and always have been directed by scientists, and there have never been any firings of directors or scientists for what appear to be political reasons.

When the Reagan administration took office, the Director was fired and the Institute's budget was severely cut. Nothing like that has ever happened to any of the other National Laboratories and Institutes.

AND MORE POLITICS

This mixing of science and politics is a dangerous tendency more suited to a "Banana Republic" than to a nation so heavily dependent on technology. Unfortunately, nuclear science has not been immune to it. Liberal Democrats seem to be against nuclear power while conservative Republicans seem to favor it. This split does not apply to the involved scientists, most of whom are politically liberal Democrats, but on the basis of their scientific knowledge, strongly favor nuclear power. I personally have been a liberal Democrat nearly all of my life, as has every member of my family for over 50 years. I, as well as the majority of my scientific colleagues are passionately devoted to the welfare of the common man (this is my favorite definition of a "liberal"). It is clear to us that his welfare is heavily dependent on a flourishing nuclear power program.

Liberal Democrat politicians have not always opposed nuclear power. These days people view nuclear power as being favored by the utilities, with lukewarm support from government bureaucrats, against the resistance of an unwilling body politic and Congress. But in the past, it was quite the opposite.[29] Congress, under the leadership of its House–Senate Joint Committee on Atomic Energy, was the original driving force behind development of nuclear power in the 1950s. The government bureaucracy represented by the Atomic Energy Commission, favored a slow, drawn-out development program, while the utility industry was heavily resistive. In fact the utilities were brought into the program only with the threat that if they didn't cooperate, the government would develop its own nuclear power program to compete with their coal- and oil-fired plants. The early history of nuclear power development was punctuated by strong pressures from Congress to speed things up. During this period and up to the early 1970s, such well-known liberal Democratic Senators as John Pastori, Clinton Anderson, Henry Jackson, Albert Gore, and Stuart Symington served on the Joint Committee and played key roles in promoting nuclear power. When information critical of it appeared on the scene, the Joint Committee held hearings in which they called in prominent scientists to explain the facts. Intelligent deliberations were held among well-informed and mutually respecting people, and questions were thereby settled.

The opposition to nuclear power among politicians started in the early 1970s when they stopped taking advice from the scientific establishment and instead began taking it from political activist groups belonging to what is generally referred to as "the environmental movement." These groups, principally led by Ralph Nader, used or desired little scientific information. They largely formulated their beliefs on the basis of political philosophy, and then found scientists wherever they could to support them, without regard to the consensus of opinions throughout the scientific community.

Somehow their political philosophy told them that nuclear energy is bad and solar energy is good. They built a case to support that position as lawyers often build cases—the only object was to win. Facts and scientific information were introduced or ignored in accordance with how they affected their case.

They weren't seeking the truth—they were sure they knew it. They were only seeking material to help them convince the public.

They sold their case to many liberals. By now "solar is good, nuclear is bad" is an article of faith in the liberal establishment and to much of the media which has strong sympathies with it. With the media on their side it can only be a matter of time before the public is won over.

In this process the involved scientific community has been essentially ignored. We are inexperienced and inept at political or media gamesmanship. We have fought, and will continue to fight, to get our message to the public. We will almost surely lose.

The problem is that our case is based on science, while the antinuclear case is based on political philosophy. When a nation whose welfare is highly dependent on technology makes vital technological decisions on the basis of political philosophy rather than on the basis of science, it is in mortal danger.

REFERENCE NOTES

1. D. Redfield, "Photovoltaics—an Overview," in "Solar Energy in Review," Solar Energy Research Institute (1981). In cases where this paper gives cost ranges, the median is used here.
2. J. L. Smith, "Photovoltaics," *Science,* **212,** 1472 (June 26, 1981).
3. E. L. Burgess, H. N. Post, and T. S. Key, "Subsystem Engineering and Development of Grid-Connected Photovoltaic Systems," Proceedings of the Fourth European Com. PV Solar Energy Conf., Stresa, Italy, May (1982).
4. Paul Maycock (U.S. Department of Energy), private communication (1982).
5. P. Masser, "Low Cost Structures for Photovoltaic Arrays," Sandia National Lab Report SAND 79-7006 (1979).
6. R. Stromberg (Sandia National Lab), private communication (1981).
7. R. Whittaker, "Photovoltaics: a Question of Efficiency," *EPRI Journal,* December (1981).
8. Pittsburgh Builder's Exchange, private communication (1982).
9. J. L. Smith, Photovoltaics as a Terrestrial Energy Source, U.S. Department of Energy Report DOE/ET-20356-6 (1980).
10. Energy Information Administration, Projected Costs of Electricity from Nuclear and Coal-fired Power Plants, U.S. Department of Energy Document DOE/EIA-0356/1, August (1982).
11. A. Szeless, "Performance of Nuclear Power Plants in 1981," *Power Engineering,* September (1982), p.54.
12. *Journal Official des Debats de l'Assemblee National,* May 10 (1982).
13. M. C. Russell, "An Apprentice's Guide to Photovoltaics," *Solar Age,* June (1981).

14. International Union of Producers and Distributors of Electrical Energy, Conference in Brussels, June 1982. Reported in *Nuclear News,* July (1982), p.48.
15. D. Van Atta, "Solar-Thermal Electric: Focal Point for the Desert Sun," *EPRI Journal,* December (1981), p.37.
16. M. L. Smith, "Wind: Prototypes on the Landscape," *EPRI Journal,* December (1981), p.27.
17. H. Neustadter (Lewis Research Center, Cleveland, Ohio), private communication (1981).
18. W. D. Metz, "Wind Energy," *Science, 197,* 971 (2 September 1977).
19. "Going With the Wind," *EPRI Journal,* March (1980), p.6.
20. "Sun-Wise Dealer Book, Appendix B: Mean Solar Radiation Maps," Sun-Wise, Inc., Great Falls, Montana (1979).
21. Calculation by author based on prices and specifications in Sears-Roebuck Catalog, assuming a house uses 200 ampere-hours of 115-volt electricity overnight.
22. J. M. Fowler, *Energy and the Environment* (McGraw-Hill, New York, 1975).
23. A. B. Lovins, "Energy Strategy: The Road Not Taken," *Foreign Affairs,* October (1976), p.65.
24. See, for example, "A Series of Critical Essays on Amory Lovins' 'Energy Strategy: The Road Not Taken,' " *Electric Perspectives,* 77/3 (1977).
25. J. A. Pickrill, C. H. Hobbs, and B. V. Mokler, "Biomedical Research Needs Related to Photovoltaic Conversion of Solar Energy to Electricity," Lovelace Research Inst. Report LF-65, U.C.-48, Albuquerque, New Mexico (1979).
26. P. D. Moskowitz, P. Perry, and I. Wilenitz, "Photovoltaic Energy Systems: Environmental Concerns and Control Technology Needs," Brookhaven National Lab Report BNL 31478 (1981).
27. R. G. Cuddihy, W. C. Griffith, C. R. Clark, and R. O. McClellan, "Potential Health and Environmental Effects of Light Duty Diesel Vehicles," Lovelace Research Institute Report LMF-89, UC-48, Albuquerque, New Mexico (1981).
28. K. A. Lawrence, "A Review of the Environmental Effects of Three Solar Technologies," International Solar Energy Society, Denver, Colorado (1978).
29. F. G. Davison, *Nuclear Power: Development and Management of a Technology* (University of Washington Press, Seattle, 1976).

Chapter 10 / WHAT THE POLLS TELL US

Some have suggested that the public is biased against nuclear energy because that technology was introduced to the world through bombs. However, there was no such antipathy toward nuclear energy in the 1950s when the major association of nuclear energy *was* bombs, or even in the 1960s when reactors began sprouting all over the country. Communities then took pride in having nuclear plants, vying for them with subtle subsidies. A 1956 Gallup poll found that only 20% of the public was opposed to having nuclear power plants in their communities. In a 1977 Roper poll, that number was 33% and in 1979 it was 56%.[1] The last rise was partly caused by the Three Mile Island accident, but the 1956 Windscale accident in England* was much more serious, and had no such effect either in England or in the United States.

---

* In the 1956 accident in Windscale, England, the fuel in the reactor caught fire, causing a considerable quantity of radioactivity to be released into the environment. That reactor was of a very different type than the ones used in the United States; such an accident would be impossible in the latter because the fuel is surrounded by water rather than by gas as in the early British reactors.

It seems clear to me that public opposition to nuclear power is something rather new and growing. In the 1960s nuclear energy was widely regarded as the energy source for the future, but in a 1980 Gallup survey of the top three preferred energy sources, 66% included solar and 36% included coal, while only 27% included nuclear.[1]

Polls of the public tell us that nuclear energy is slipping badly. But what about polls of the scientific community? How do scientists feel about these issues?

ROTHMAN–LICHTER POLLS

In 1980, Stanley Rothman and Robert Lichter, social scientists from Smith College and Columbia University, respectively, conducted a poll[2] of a random sample of scientists listed in *American Men and Women of Sciences,* the "Who's Who" of scientists. They received a total of 741 replies. They categorized 249 of these respondents as "energy experts" based on their specializing in energy-related fields rather broadly defined to include such disciplines as atmospheric chemistry, solar energy, conservation, and ecology. They also categorized 72 as nuclear scientists based on fields of specialization ranging from radiation genetics to reactor physics.

Some of their results are listed in Table 1, from which we see that 89% of all scientists, 95% of scientists involved in energy related fields, and 100% of radiation and nuclear scientists favor proceeding with the development of nuclear power. Incidentally, there were no significant differences between responses from those employed by industry, government, and universities. There was also no difference between those who had, and had not, received financial support from industry or the government.

Another interesting question was whether they would be willing to locate nuclear plants in cities in which they live (actually no nuclear plants are built within 20 miles of the boundaries of large cities). The percentage saying that they were willing was 69% for all scientists, 80% for those in energy-related sciences, and 98% for radiation and nuclear scientists. This is in direct contrast to the 56% *opposition* from the general public.

Rothman and Lichter also surveyed opinions of various categories of media journalists, and developed ratings for their support of nuclear energy. Their results are shown in Table 2.

We see clearly that scientists are much more supportive of nuclear energy than are journalists. Even among journalists, science journalists are much more supportive than journalists in general; the latter, for some reason, author most of the media material on nuclear energy. The most influential journalists,

TABLE 1
HOW SHOULD WE PROCEED WITH POWER DEVELOPMENT?
RESULTS FROM REF. 2

	All scientists (%)	Energy experts (%)	Nuclear experts (%)
Proceed rapidly	53	70	92
Proceed slowly	36	25	8
Halt development	7	4	0
Dismantle plants	3	1	0

those from the top newspapers and TV networks, are less favorable to nuclear energy than their colleagues. These findings do much to explain the problems with media coverage discussed frequently in this book.

BATTELLE AND MEDIA INSTITUTE STUDIES

The Battelle Memorial Institute, a nonprofit foundation, did a count of pro- and antinuclear articles in four national periodicals including the *New York Times*.[3] It found that in 1972 slightly more pro- than antinuclear arti-

TABLE 2
SUPPORT FOR NUCLEAR ENERGY[a]

Category	Number surveyed	Support rating
Nuclear scientists	72	7.9
Energy scientists	279	5.1
All scientists	741	3.3
Science journalists	42	1.3
Prestige press journalists	150	1.2
Science journalists at *New York Times, Washington Post,* TV networks	15	0.5
TV reporters, producers	18	−1.9
TV journalists	24	−3.3

[a] Scale runs from +10 for perfect support to −10 for complete rejection. From Ref. 2.

cles appeared, but in 1976, negative articles outnumbered positive ones by 2 to 1.

The Media Institute, a Washington-based organization, published an extensive study of TV network evening news broadcasts during the 1970s.[4] By far the most often quoted source of information was the antinuclear activist organization Union of Concerned Scientists whose estimated[2] membership includes only about 0.1% of all scientists. The most widely quoted "nuclear expert" was Ralph Nader. During the month following the Three Mile Island accident, the only scientist quoted was Ernest Sternglass, almost universally regarded in the scientific community (including antinuclear scientists) as one of the least reliable of all scientists in the field of radiation health (cf. below). From May 1978 through April 1979, there were no pro-nuclear "outside experts" among the ten top quoted sources. The only pro-nuclear sources in the top ten were from the nuclear industry, who clearly had low public credibility, especially during that period.

A POLL OF RADIATION HEALTH SCIENTISTS

After completing the first draft of this book, I found myself bothered by some disturbing questions. I was claiming that the opinions expressed in the book were those of the great majority of the involved scientific community, but how could I prove that to the readers? In fact, how could I be sure of it myself? My impressions were based on innumerable conversations, remarks, and innuendos, but the total number of people encompassed by these was only a tiny fraction of the total community. Moreover, people may not say publicly what they really believe because of various pressures they may feel.

In order to resolve some of these questions, I decided to conduct a mail poll to determine the opinions of the community of radiation health scientists. The principal professional societies serving this community are the Health Physics Society (HPS) and the Radiation Research Society (RRS). Memberships are reasonably inexpensive (and tax deductible) and they include free subscriptions to their journals *Health Physics* and *Radiation Research,* which publish many of the scientific papers workers in this field would have to read; for these reasons and others, nearly all radiation health scientists belong to one or both. Both publish membership lists giving names and addresses, between the two a total of 5000. I decided to include only those employed by universities, which reduces the suspicion that their opinions are influenced by the security of their employment and selects those more likely to be close to research. Since limiting the study to the university-employed still did not sufficiently reduce the total number to be polled, a random sampling of these was used.

The membership lists are in alphabetical order and published on about 122 pages and 40 pages with two columns to a page for HPS and RRS, respectively. My random selection process was to choose the last university-employed member in each column from the HPS list and the first and last from the RRS list. This gave 310 people to be polled with roughly equal numbers from the two societies (some of the columns in the HPS list had no university-employed people, and due to a misunderstanding with my student assistant, the last 14 pages of the HPS list were not used).

The letter sent to each of these, copied verbatim (with the exception of question 5 to be discussed below), was the following:

> I am conducting an opinion poll on some important topics of public interest, of randomly selected University connected members of Health Physics Society (back sheet, white) and Radiation Research Society (back sheet, yellow). I would be very appreciative if you could reply to the questions posed below and return the form in the enclosed stamped and addressed envelope within the next few days. It should require only a very few minutes of your time. You may note that your reply is completely anonymous; there is no way to trace how anyone replied, and I promise that no attempt will be made to do so. If there are questions which you prefer not to reply to, simply leave them blank.
>
> Sincerely yours,
>
> Bernard L. Cohen

QUESTIONNAIRE

1. In comparing the general public's fear of radiation with the actual dangers of radiation, I would say that the public's fear is (check one):

2	grossly less than realistic (i.e., not enough fear)
9	substantially less than realistic
8	approximately realistic
18	slightly greater than realistic
104	substantially greater than realistic
70	grossly greater than realistic (i.e., too much fear)
211	Total

2. The impressions created by television coverage of the dangers of radiation (check one):

 _59_   grossly exaggerate the danger
 110   substantially exaggerate the danger
 _26_   slightly exaggerate the danger
 __5_   are approximately correct
 __3_   slightly underplay the danger
 __2_   substantially underplay the danger
 __1_   grossly underplay the danger
 206   Total

3. The impressions created by newspaper and popular magazine coverage of the dangers of radiation (check one):

 _47_   grossly exaggerate the danger
 113   substantially exaggerate the danger
 _38_   slightly exaggerate the danger
 __8_   are approximately correct
 __3_   slightly underplay the danger
 __2_   substantially underplay the danger
 __0_   grossly underplay the danger
 211   Total

4. From the standpoint of our national welfare and in comparison with other health threats from which the public needs protection, the amount of money now being spent on radiation protection in the United States is (check one):

 _18_   grossly excessive
 _35_   substantially excessive
 _30_   slightly excessive
 _62_   about right
 _22_   slightly insufficient
 _21_   substantially insufficient
 __4_   grossly insufficient
 192   Total

5. How would you rate the scientific credibility in the field of radiation health of the following scientists or groups of scientists judging only by their work and public statements since 1972 (i.e., ignoring everything they did before that time)? In *each* space, write a number between 0 and

100 indicating in which percentile of credibility of radiation health scientists their credibility falls; for example, "60" would mean that that person's work is more credible than that of 60% of all scientists in the field, and less credible than that of 40%. All of these have spoken or written on the radiation controversy and many scientists feel that anyone who engages in public debate compromises his credibility; please try to ignore this prejudice. If you are not familiar with the person's work or statements, leave blank.

E-82(175)	BEIR Committee (National Academy of Sciences)
E-81(116)	UNSCEAR Committee (United Nations Scientific Committee)
A-25(157)	The Ralph Nader research organizations
E-85(157)	ICRP
A(?)-41(122)	Union of Concerned Scientists
E-85(157)	NCRP
E-72(60)	WASH-1400 Panel on Health Effects
E-78(102)	Establishment Scientist A (institutional connection)
A-14(154)	Ernest Sternglass (University of Pittsburgh)
E-71(113)	Establishment Scientist B (institutional connection)
A-30(129)	John Gofman (Livermore Laboratories and University of California—retired)
E-75(52)	Establishment Scientist C
A-28(96)	Thomas Mancuso (University of Pittsburgh)
E-80(114)	Establishment Scientist D
A-32(60)	Thomas Najarian—studied Portsmouth, New Hampshire shipyard workers
E-71(80)	Establishment Scientist E
A-19(52)	Irwin Bross (Roswell Park Memorial Institute)
E-81(38)	Establishment Scientist F
A-26(107)	Helen Caldicott—Physicians for Social Responsibility

6. In comparing the dangers of radiation from a full nuclear power industry (at least 50% of all electricity generated by nuclear power plants) with the dangers of air pollution from a full coal-powered industry, including accidents as well as routine operation in both cases, which of the following represents your views?

   7   nuclear is grossly more dangerous
  10   nuclear is substantially more dangerous
  21   nuclear is slightly more dangerous
  31   the two are roughly equally dangerous

   <u>40</u> coal is slightly more dangerous
   <u>79</u> coal is substantially more dangerous
   <u>15</u> coal is grossly more dangerous
  203 Total

7. The news media have frequently charged or implied that United States
   Government agencies have been systematically suppressing unfavorable
   information about the dangers of radiation. In comparison with how
   government agencies treat other matters, I consider their *openness* and
   honesty (opposite of suppression) with regard to radiation dangers to have
   been (check one reply each for prior to 1972 and since 1972; leave blank
   if you have no basis for an opinion).

prior to 1972   *since 1972*
   <u>2</u>   <u>17</u>   far above average (do *not* suppress)
   <u>23</u>   <u>78</u>   above average
   <u>33</u>   <u>70</u>   about average
   <u>74</u>   <u>23</u>   below average
   <u>36</u>   <u>1</u>   far below average (do suppress)

8. My highest academic degree is (check one)

   <u>156</u> doctorate
    <u>41</u> master's
    <u>10</u> bachelor's
     <u>3</u> less than bachelor's

In the spaces left for the respondent to mark his selection, I have placed the
totals from the 210 responses received (68% of the questionnaires sent out).
There was little difference between the replies from the two societies. For all
questions, there were at least a few respondents who chose not to vote, but
the number so choosing was especially large on question 5; the results quoted
there are the *average* of all responses, followed in parentheses by the number
of responses received. I have also added an E or an A, indicating whether
the person named supports "The Establishment"—favoring the views of BEIR,
UNSCEAR, ICRP, and NCRP, the views presented in this book—or "anti-
Establishment"—claiming that radiation is more dangerous than it is judged
to be by "The Establishment."

     If anyone would like to repeat the poll, using the same or different
random selection procedures, I offer to pay double the cost of the process
(about $300) if his results are appreciably different from mine; or if some

organization widely recognized to be neutral is willing to repeat it, I will pay the cost regardless of the outcome. To me, a very important property of this poll is that the replies are completely untraceable, so no respondent had any reason not to express his feelings freely.

Each reader is free to interpret the results in his own way. In the following discussion, I offer my interpretations.

Question 1

The poll strongly favors the position that the public's fear is "substantially" or "grossly" greater than the actual dangers of radiation. The favoring of "substantially" over "grossly," is in my view probably due to the fact that scientists traditionally shy away from use of superlatives.

The 11 respondents who said that the public's fear is *less* than realistic might appear to support the position that radiation is much more dangerous than estimated by "the Establishment." However, in question 5, seven of these eleven rated *all* of the principal Establishment groups—BEIR, UNSCEAR, ICRP, and NCRP—higher than *any* of the antiestablishment scientists, and in only two of the other four was there an antiestablishment pattern in the reply to question 5 or the other questions (one of these two did not reply to question 5). The first mentioned seven apparently believe that the public is very unconcerned about the dangers of radiation.

The poll does show, however, that 82.5% of the involved scientists believe that the public's fear is substantially or grossly excessive. That is the principal point to be made here.

Question 2

The results of the poll favor the position that television coverage "substantially" or "grossly" exaggerates the danger of radiation. Again the favoring of "substantially" over "grossly" may be due to the aforementioned aversion to superlatives. Only 3 of the 6 who say that TV underplays the dangers of radiation are among those who say (in question 1) that the public's fear is less than realistic. Of the other 3, none show an antiestablishment attitude in question 5. It is perhaps refreshing to mention that a few said that they could not reply to question 2 because they never watch television.

At any rate, the principal point to be made here is that 82% of the respondents say that the impressions created by TV coverage substantially or grossly exaggerate the danger.

264 | CHAPTER 10

Question 3

The results here, on newspaper coverage, are quite similar to those in question 2 on TV coverage, although there is a slightly lower percentage, 76% versus 82%, who say that the danger is substantially or grossly exaggerated. I would agree that newspaper coverage has been slightly more balanced than TV. All five of those who say that newspapers underplay the danger are among the six who say in question 2 that TV underplays the danger.

Question 4

This question was inserted to test the claim of antinuclear activists that scientists and engineers favor nuclear power because it benefits their pocketbooks. If that were so, they would reply that the money being spent on these problems is insufficient. However, the results show the opposite to be true; only 24% say that it is insufficient, whereas 43% say that it is excessive and 28% say that this excess is substantial or gross. I am certain that in any other field of science, there would be a heavy majority claiming that funding should be increased. It is only natural to believe that the research you are doing is very important.

Question 5

In formulating this question, proestablishment and antiestablishment people and groups were intentionally intermixed so as not to bias respondents by the positioning. The major establishment groups, BEIR, UNSCEAR, ICRP, and NCRP all received average ratings in the range 81–85. Of the 175 ratings for BEIR, only one was below 50; the number of ratings below 50 for the others was 3 for UNSCEAR, 0 for ICRP, and 2 for NCRP, with only one of these six below 40.

The Ralph Nader research organizations often get media coverage, but they have very low credibility among scientists, with an average rating of only 25. Only 4 of the 157 ratings for it were above 75, and only one was above 85. The Union of Concerned Scientists has not been antiestablishment on scientific questions about the dangers of radiation,* and in fact accepts

---

* The scientists whose work has been criticized in this book—Sternglass, Gofman, Caldicott, Mancuso, Bross, and Najarian—are not associated with Union of Concerned Scientists (UCS), and UCS has never supported their claims. Virtually all of the thrust of UCS activities has been on the probability of reactor accidents. This is a very complex subject not generally familiar to radiation health scientists. Note that I have given UCS estimates in many places in this book.

and uses the BEIR recommendations. I therefore labeled them as A(?). Its rating of 41 is considerably higher than that of the Nader groups, but still very far below that of the establishment groups.

For reporting the results of this poll, those people who have been identified have injected themselves into the forefront of this controversy. Those not named have not sought to publicize their stand on this issue. (There were also names of two other scientists on the questionnaire, that fit in neither category, but I included them to satisfy my personal curiosity. Their results are not listed.) All names appeared on the original ballot.

The ballot was prepared, presented, and tallied in a strictly neutral manner. Ordering of names was random except for a conscious intermingling of establishment and antiestablishment names. The establishment scientists all received average credibility ratings between 71 and 81, whereas those whose work has been criticized here all received average ratings of 32 or less, a dramatic difference. The lowest average rating, 14, was for Ernest Sternglass, who had only 3 ratings out of 154 above 65. The highest average rating among this group, 32, was for Najarian, who later withdrew most of his claims.

I interpret the replies to question 5 to indicate that the radiation health scientific community strongly supports the establishment and has little sympathy with the antiestablishment scientists.

Question 6

The results show that 66% of the respondents believe that coal is more dangerous than nuclear energy while only 19% believe that nuclear energy is more dangerous. Radiation health scientists are not generally knowledgeable about air pollution problems. I therefore do not feel that this negates in any way the unanimous conclusion of the many studies cited in Chapter 1 that coal is more dangerous than nuclear energy. On the other hand it does confirm that radiation scientists do not consider the radiation risks of nuclear power, with which they are intimately familiar, to be as dangerous as the source of electricity we have been living with throughout our lives.

Question 7

The consensus seems to be that prior to 1972 the government, which was then represented largely by the old Atomic Energy Commission, was less open and honest than average, but that the situation has been reversed since 1972 with the Department of Energy and the Nuclear Regulatory Commission being above average in openness and honesty. I had no personal experience in these matters before 1972 when I first became involved in nuclear power issues; I therefore cannot express an opinion on the earlier period. However, I have called those agencies hundreds of times in recent years in seeking technical information, and they have always been quite forthcoming. This is a matter any citizen can check for him or herself if he or she is looking for an answer to a technical question. Simply call, ask for someone involved in the technical area of interest, introduce yourself, and ask your question.

SUMMARY

If nothing else, the results of this poll show a pattern of what radiation health scientists believe that is very different from the conception portrayed by the media and accepted by the general public. They believe that the public's fear of radiation is greatly exaggerated, and that media coverage greatly exaggerates the danger. They strongly support the establishment committees and the scientists on them, and give very little credence to the antiestablishment scientists who have received so much media coverage.

REFERENCE NOTES

1. R. Kasperson, G. Berk, A. Sharaf, D. Pijawka, and J. Wood, "Public Opinion and Nuclear Energy: Retrospect and Prospect," *Science, Technology, and Human Values,* Spring (1980), p.11.
2. S. Rothman and S. R. Lichter, "The Nuclear Energy Debate: Scientists, the Media, and the Public," *Public Opinion,* August–September (1982), p.47.
3. S. Nealy and W. Rankin, "A Comparative Analysis of Print Media Coverage of Nuclear Power and Coal Issues," Battelle Human Affairs Research Center, Seattle (1979).
4. The Media Institute, "Television Evening News Covers Nuclear Energy," Washington, D.C. (1979).

Chapter 11 / QUESTIONS FROM THE AUDIENCE

In teaching classes and lecturing, there are always questions from the audience. In this chapter we consider a representative group of such questions.

RADIOACTIVITY AND RADIATION

Q: Radioactivity can harm us by radiating us from sources outside our bodies, by being taken in with food or water, or by being inhaled into our lungs, but only one of these pathways has been considered in your lecture. What about the others?

A: All of these pathways are treated in all cases in the scientific literature, but in order to simplify popular expositions it is usual to treat only the most important pathway. With buried radioactive waste, for example, it is usual to discuss only contamination of food and water. Since the most important escape mechanism is becoming dissolved in groundwater, it seems evident that food or water contamination would give it the best chance of getting into human bodies. Elaborate calculations confirm this.

Q: Cancers from radiation may take up to 50 years to develop, and genetic effects may not show up for hundreds of years. How, then, can we say that there will be essentially no health effects from the Three Mile Island accident?

A: We have measured the radiation doses in millirem, and from the experience and scientific research outlined in Chapter 2, we know what the eventual health impacts from a given radiation dose will be.

Q: Measurements of radioactivity in air, for example, are made at a few monitoring stations. How do we know the levels may not be much higher at places where there are no monitoring stations?

A: Scientists choose locations of monitoring stations so as to minimize this possibility, making use of the considerable body of scientific information on how materials are dispersed under various weather conditions. This information also predicts relationships between readings at various stations, which are checked to give added confidence. Radiation levels at various locations can be predicted from the quantity of radioactivity released and a knowledge of the weather conditions including wind speed and directions, and temperature versus height above ground. The weather conditions around nuclear plants are constantly monitored. Much can be learned about a radioactivity release from measuring radioactivity on the ground surface up to several days later. In the Three Mile Island accident, air samples were collected by airplanes to give additional data, and photographic film from area stores was purchased and developed to measure the fogging by radiation. (None was observed, but it would have been if there had been appreciable radiation.) There are thus many independent ways to determine the pattern of radiation exposure, and they serve as checks on one another.

This situation contrasts sharply with that for air pollution from coal burning. Since monitoring for air pollutants is much more difficult and expensive, there are very few monitoring stations even in a large metropolitan area. Nonetheless air pollution kills many thousands of people every year and is thus a much greater threat to our health than is radioactivity from nuclear plants.

Q: Radioactive materials can be concentrated by various biological organisms. For example, strontium-90 ingested by a cow mostly gets into its milk. Doesn't this make radioactivity much more dangerous than your calculations indicate?

A: This is taken into account in all calculations and estimates. There was a widely publicized omission for the case of strontium-90 in milk in the 1950s, but that was a very long time ago, scientifically speaking.

Q: Air pollution may kill people now, but radiation induces genetic effects that will damage future generations. How can we justify our enjoying the benefits of nuclear energy while future generations bear the suffering from it?

A: Air pollution and chemicals released in coal burning also have genetic effects as indicated by tests on microorganisms. While they are not as well understood and quantified, there is no reason to believe that the genetic effects of coal burning are less severe or fewer than those of nuclear power.

The genetic impacts of radiation are not large. The total number of eventual genetic defects caused by a given radiation exposure added up over all future generations is less than the number of cancers it causes.

There are many ways in which our technology injures future generations, such as consuming the world's limited mineral resources, which will cause them infinitely more serious injury than genetic effects of our radiation. In the latter case, we are more than compensating our progeny with biomedical research which will greatly improve their health in many ways, including averting much of their genetic disease. (Genetic effects are discussed extensively in Chapter 2.)

Q: Can the genetic effects of low-level radiation destroy the human race?

A: No. The law of natural selection causes good mutations to be bred in and bad mutations to be bred out. In the very long term, mutations from any given amount of radiation exposure disappear if they are harmful, and are preserved if they are beneficial; they can therefore only improve the human race, although that effect is extremely small.

Mankind has always been exposed to radiation from natural sources, hundreds of times higher in average intensity than low-level radiation from the nuclear industry. Yet even this natural radiation is responsible for only a few percent of all genetic disease. Spontaneous mutations are responsible for the great majority of it.

Q: Isn't the artificial radioactivity created by the nuclear industry, to which humans have never been exposed until recently, more dangerous than the natural radiation which has always been present?

A: The cancer and genetic effects of radiation are caused by a particle of radiation, say a gamma ray, knocking loose an electron from a certain molecule. In this process, there is no possible way, even in principle, for the electron or the molecule to "know" whether that gamma ray was originally emitted from a naturally radioactive atom, or from an atom that was made radioactive by nuclear technology. The answer to the question is NO.

Q: Can radiation exposure to parents cause children to be born with two heads or other such deformities?

A: NO. Such things are not occurring now although mankind has always been exposed to natural radiation. There is no possible way in which artificial radiation can cause problems that do not occur as a result of natural radiation.

Q: Is there any factual basis for radiation creating monsters like "The Incredible Hulk"?

A: Absolutely none. These are strictly creations of an artist's imagination.

Q: You frequently use statistics to support your case. But it is well known that, "while statistics don't lie, liars can use statistics." How can we trust your statistics?

A: I make very little use of statistics because there is no statistical evidence of harm to human health due to radiation from the nuclear industry. I use probabilities, which are something very different.* If you don't believe in probability, I would love to engage you in a game of coin flipping or dice rolling. Las Vegas, the various state lotteries, and insurance companies, are doing very well depending on the laws of probability.

Antinuclear activists are the ones who are forever trying to use statistics to deceive. If you look hard enough, you can always find an area around some nuclear plant that has a higher than average cancer rate. Of course you can just as easily find one with a lower than average cancer rate, but they never bother to report that.

Q: How can you treat deaths due to radiation as statistics? These are human beings suffering and dying.

---

* To illustrate the difference between probability and statistics, consider the honest flipping of a coin. The *probability* for heads is 50%. If one flips a coin ten times and gets eight heads, he could say that his statistics indicate that heads comes up 80% of the time.

A: There are also human beings suffering and dying from air pollution, from chemical poisons, from poverty, etc. Nuclear power will reduce these problems. I am only interested in reducing the total number of people who suffer and die.

Q: Isn't a nuclear accident that kills 1000 people worse than having 10,000 people die one by one from air pollution with no one knowing why they died?

A: I thoroughly disagree. The only reason anyone could believe such a thing is because the media handles it that way. They would give tremendous publicity to a nuclear accident killing a thousand people (or even a hundred people) but they hardly mention the 10,000 or so people who die from air pollution every year. But we must not be brain-washed by that non-sense—we must recognize that the media are basically in the entertainment business.

If you would choose a technology that kills 10,000 per year with air pollution over one that kills 1000 in a large accident each year, you should be given the job of explaining to the extra 9000 victims (and their loved ones) that they must die because people don't like media reports of large accidents. I'm sure you would quickly change your mind.

Q: Does radiation make people glow in the dark?

A: Radiation produces light only when extremely intense, as inside a nuclear reactor; a person exposed to that much radiation would die instantly (as he would inside any other furnace). Comedians seem to get laughs with jokes about people glowing in the dark from radiation, but there is no factual basis for that idea.

Q: What does it feel like to be exposed to radiation?

A: We have all experienced that in getting medical or dental X-rays which give much higher exposures than anyone gets from the nuclear industry. People treated with radiation therapy get very high doses, millions of millirem. The answer is that there is no feeling or sensation from radiation exposure.

Q: If we can't feel radiation, how do we know when we are being exposed?

A: There are numerous instruments for detecting radiation. Many of them are very cheap, reliable, and sensitive. It is very much easier to detect radiation than the dangerous components in air pollution.

TRUST AND FAITH

Q: Why should we believe scientists when they have made nuclear bombs and all sorts of devastating weapons?

A: Working for the military is not an indication that a person does not tell the truth. Moreover, the scientists involved with nuclear power are an entirely different group (with a very few individual exceptions) from those who developed nuclear weapons.

 I have never heard evidence that the scientific community has deceived the public. That community is so diverse that it would be impossible for it to do so. Moreover, working as a scientist requires a high degree of honesty, if for no other reason than that dishonesty would be readily discovered and the career of an offender would be irreparably damaged by it.

Q: Since nuclear scientists rely on the nuclear industry for their livelihood, how can we believe them?

A: *University* scientists do not rely on the nuclear industry; in fact most of them have lifetime job security guaranteed by the university that employs them (I am in that position). Radiation health scientists, would get *increased* importance and job security if people decided that radiation is *more* dangerous, because they are the ones who protect the public from radiation. If the nuclear industry were to shut down today, they would have a secure lifetime career from participating in the retirement of plants.

 Antinuclear activist scientists, on the other hand, generally make a good living out of their opposition to nuclear power. They get large fees for speaking and their books sell well. If a nuclear scientist were interested in making extra money, he would do well to reverse his position and become antinuclear.

 A university nuclear scientist could become antinuclear without any repercussions to his job security. The same is largely true for those employed by the government. Scientists employed by antinuclear activist organizations, on the other hand, would instantly lose their jobs if they decided to become pro-nuclear.

Q: With nuclear scientists split on the question of dangers of radiation, how do we know which side to believe?

A: The split in the scientific community is not "down the middle" as the media would have you believe, but is very heavily one-sided. All of the official committees of prestigious scientists, BEIR, ICRP, UNSCEAR,

NCRP, and many others in foreign countries agree unanimously on the effects of radiation (within a range of differences that is irrelevant for purposes of public concern), and they are backed by the vast majority of the involved scientific community. There has not been even a single vote by a single scientist on these committees supporting the views of anti-nuclear activists, the views so widely trumpeted by the media. The "split" is largely a media creation (cf. Chapter 10).

Q: Why should we trust the government when it gives us information on nuclear power?

A: The scientific evidence on health impacts of radiation has little dependence on government sources of information. The same is true of most other areas covered in this book. You are mainly being asked to trust the international scientific community.

　　When it comes to questions of whom you can trust, surely the media must rank near the bottom of the list. Many cases have been cited in this book where they have knowingly deceived the public. Inadvertent deception, caused by political prejudice or simple ignorance, is an everyday media practice. Their priority is capturing and holding an audience, not dispensing correct information. Even with the best of intentions, it is difficult for a media person to report on things he doesn't understand himself. Yet the public derives most of its information on nuclear power from the media.

Q: The nuclear establishment told us that there could never be a reactor accident, but we had Three Mile Island. How can we trust them?

A: The nuclear establishment did not say that there could never be a reactor accident. The Rasmussen Study, which represents the nuclear establishment more than anything else on that issue, estimates that there is a 20% chance that we would have had a meltdown by now (between civilian and naval reactors) whereas there has been none. It predicts that we should have had two loss-of-coolant accidents by now, whereas Three Mile Island has been the only one.

Q: How can we trust the nuclear establishment when they construct nuclear power plants on earthquake faults?

A: The closest distance from any nuclear power plant to a fault capable of causing significant damage is 3.3 miles for the Diablo Canyon power plant in California. (NRC Atomic Licensing Appeals Board Decision No. 644, June 16, 1981); the basis for the question is, therefore, exaggerated. The regulations on location of reactors relative to earthquake faults are very lengthy, complex, and technical. They have been worked out by some of the nation's foremost earthquake scientists. The distance from a fault allowed depends on the type of fault and the length of time since it has been active. I personally have confidence that these earthquake experts know what they are doing, or at least know a lot more than the propaganda artists who delight in attacking their decisions.

Q: How can we trust reactor operators to do their job properly? How do we know they won't get drunk and cause an accident?

A: In safety analyses, it is not assumed that reactor operators are perfect; it is rather assumed that they make mistakes just like anyone else, pushing wrong buttons, failing to do required jobs, etc. In accordance with the principle of defense in depth, the guiding philosophy in power plant design, there are back-up systems to compensate for such errors. The failure of any one link, of course, reduces the effectiveness of the defense in depth. It is therefore avoided as much as possible. Reactor operators must frequently pass stiff examinations; there are several operators on hand at all times; there are training programs, supervision, and inspections. But it is clearly recognized by all concerned that reactor operators are human beings and must be expected to behave as other human beings, which is a long way from perfect.

Q: How can we trust utilities not to take shortcuts in efforts to save money, thereby compromising safety?

A: Since I have no firsthand experience with utility construction practices, I cannot speak as an expert on this. But utilities are guaranteed a reasonable profit by their Public Utility Commissions if they behave properly. They therefore have no great incentive to save money by cutting corners. Moreover, an accident in a nuclear plant is perhaps the most serious business blow a utility can suffer, costing its stock holders hundreds of millions or even billions of dollars. The utility that owns the Three Mile Island plant is nearly bankrupt as a result of that accident.

In addition the Nuclear Regulatory Commission (NRC) makes regular inspections. Anything not properly constructed may have to be torn out and reconstructed, at very great expense. A plant near Cincinnati is in deep trouble because of this.

There is therefore a heavy incentive for utilities to behave properly in constructing plants. Similar considerations apply to operating plants. They are inspected frequently by NRC, and heavy fines are levied for substandard practices. There are also careful inspections by the Institute for Nuclear Power Operations. This is an Atlanta-based organization sponsored by the nuclear industry because it recognizes that an accident in one plant causes difficulties for the whole industry.

REACTOR ACCIDENTS AND SAFETY

Q: Can a reactor explode like a nuclear bomb?

A: No, this would be impossible. A bomb requires fuel highly enriched in fissile material, whereas reactor fuel has only 3% fissile material. A bomb cannot work if the fuel is surrounded by water because the hydrogen in water slows down the neutrons. There are other reasons in addition that make a nuclear explosion impossible.

Q: Was the Three Mile Island accident a "close call" to disaster?

A: All studies agree that it was not (details are given in Chapter 3).

Q: Is nuclear power safe?

A: Nothing in this world is perfectly safe. But in comparison with other methods available for generating electricity, or with the risks of doing without electricity, the dangers of nuclear power are very small.

Q: Nuclear power is very new and different. How do you know that new and unsuspected problems will not develop?

A: There has been a tremendous research effort on areas of potential trouble, and commercial reactors have been operating for over 25 years. The U.S.Navy has been operating a large number of reactors for 20 years or more. But of course new and unsuspected problems have developed (cf. Chapter 3) and probably will continue to develop as in any technology. That is why there are continuing research efforts, many avenues for information exchange among plants, and continued attention to the problems by NRC. I see no reason to believe that we cannot keep ahead of the problems and maintain the present level of safety. As experience accumulates, we learn more about safety problems, thereby *improving* safety. Many valuable lessons were learned from the Three Mile Island accident.

Q: How can you trust the "fault tree analysis" method used in the Rasmussen Study?

A: It is the best method available. If you don't trust it, you can fall back on experience. We have had about 4,000 relevant reactor-years of experience, with no meltdowns, and no very "close-calls" to a meltdown. Moreover, as a result of experiences with various malfunctions, reactor safety is constantly being improved—lots of improvements were made as a result of lessons learned from the Three Mile Island accident. It therefore seems reasonable to conclude that a meltdown will not occur much more frequently than once in 4,000 reactor-years. This is not much less than what fault-tree analyses give. Don't forget—most meltdowns cause no health effects.

Q: If reactors are so safe, why don't home owner's insurance policies cover reactor accidents? Doesn't this mean that insurance companies have no confidence in them?

A: Insurance companies do insure reactors. In fact they stand to lose more money from a nuclear accident than from any other readily conceivable mishap. However, they are limited by law in the amount they can insure against any one event, because if an insurance company were to fail, many innocent policy holders would be left without protection.

Liability insurance for reactors is covered by an act of Congress which requires $560 million of no-fault insurance on each plant (this amount is scheduled to rise to nearly $1 billion), the first $160 million from private insurance companies and the remainder from a pool maintained by all nuclear power plants. Because of this insurance, coverage in home-owner's insurance would provide double coverage and is therefore excluded. If the liabilities from an accident should exceed the maximum (currently $560 million), Congress has stipulated that it will provide relief as is customary in disaster situations. Few other disasters are covered by as much private insurance. For example, there is little or no coverage for dam failures, bridge collapses, etc.

Q: If reactors are safe, why are there evacuation plans for areas around them?

A: This is an example of regulatory ratcheting by the Nuclear Regulatory Commission. Until recently, there were no such plans, and they are not used in other countries. There are no evacuation plans around chemical plants although evacuations in their vicinity are more likely to be necessary than around nuclear plants. Most evacuations occur as a result of railroad or truck accidents involving toxic chemicals, but there is no advanced planning for them. It would be difficult to dispute the NRC viewpoint that having evacuation plans increases safety to some extent. They gave no consideration to the fact that the existence and advertising of these plans is unsettling to the public.

RADIOACTIVE WASTE

Q: What are we going to do with the radioactive waste?

A: We are going to convert it into rocks, and put it in the natural habitat of rocks, deep underground.

Q: How do we know it will be safe there?

A: We know a great deal about how rocks behave, and if the waste-converted-to-rock behaves that way, it will be extremely safe (cf. evaluation in Chapter 5).

Q: If it's so easy, why aren't we burying waste now?

A: One reason is that we really don't have any waste ready to bury. The preparation of waste for burial awaits resolution of the reprocessing issue.

   Another reason is that research is being done on determining the *optimum* burial technology. While most sites would provide adequate safety, some would be safer than others. There is therefore a rather large effort on site selection.

   From an objective viewpoint, there need be no rush to bury the waste, and in fact there would be important advantages in storing it for 50–100 years before burial—this is the plan in Sweden. However, because of the intense public pressure generated by public misunderstanding, the U.S. government is forging ahead rapidly with plans to begin burying waste by the mid-1990s.

Q: Isn't disposal of radioactive waste an "unsolved problem"?

A: What is a solved problem? Some of the wastes from coal burning, better known as "air pollution," are simply spewed out into the air where they kill many thousands of Americans every year. Is that a solved problem? The solid wastes from coal burning are dumped on the ground where, over the next hundred thousand years or so, they will kill many more thousands (cf. chapter 5). Is this a solved problem?

   We know many satisfactory solutions to the radioactive waste burial problem, and are now involved in deciding which of these is best. Any one of them would be thousands of times less damaging to health than our handling of wastes from coal burning. Since we have no nuclear waste ready to bury, there is no time lost in doing further studies. Does the fact that we haven't yet chosen from among the various acceptable options make this an unsolved problem? Clearly we are dealing here with rhetoric and propaganda rather than with a substantive issue.

Q: How long will the radioactive waste be hazardous?

A: It will lose 98% of its toxicity after about 200 years, by which time it will be no more toxic than some natural minerals in the ground. It will lose 99% of its *remaining* toxicity over the next 30,000 years, but it will still retain some toxicity for millions of years.

This situation is much more favorable than for some toxic chemical agents like mercury, arsenic, cadmium, etc. which retain their toxicity undiminished *forever*.

Q: How will we get rid of reactors when their useful life is over?

A: Fuel, which contains nearly all of the radioactivity, is removed every year. When a reactor is retired, the remaining fuel will be similarly removed and sent away for reprocessing and disposal as high-level waste. The residual equipment which is only weakly radioactive becomes low-level waste. There are several proposals for how it will be handled—only a few small reactors have been dismantled so far. Studies show that paying for this process adds at most a few percent to the cost of electricity.

MISCELLANEOUS TOPICS

Q: How long will our uranium supplies last?

A: With present reactors, we can continue building plants for about another 30 years and still be able to guarantee each a lifetime supply of fuel. Beyond that we will have to convert over to breeder reactors. With that technology, our fuel supply will last forever without affecting the price of electricity (cf. Chapter 7).

Q: Is nuclear power necessary?

A: The United States could get along for the foreseeable future with coal. It would be more expensive as well as much more harmful to our health and to the environment, but we could get by. For other countries, the situation is much less favorable. Western Europe and Japan have relatively little coal, and will therefore sorely need nuclear power when the oil runs out or is withdrawn for political or economic reasons. For them, nuclear power is much cheaper than any alternative.

Q: What harm could terrorists do if they took control of a nuclear power plant?

A: In principle, they could cause a very bad accident, thereby killing tens of thousands of people, including themselves. However, nearly all of their victims would suffer no immediate effects, but rather would die of cancer 15 to 50 years later. In view of the high normal incidence of cancer, these excess cases would be unnoticeable (cf. Chapter 3).

By contrast, there are many simple ways these terrorists could kill at least as many people immediately. For example, they could put a poison gas into the ventilation system of a large building. Other examples are given in Chapter 7.

Nuclear power plants have very elaborate security measures with over a dozen armed guards on duty at all times, electronic aids for detecting intruders, emergency procedures, radio communication, etc. To sabotage a nuclear plant effectively would require a considerable amount of technical knowledge, or a truck load of explosive.

If terrorists *only* wanted to destroy the electricity generation capacity, it would be infinitely easier to sabotage a coal burning power plant. Coal plants have virtually no security measures and no sabotage resistance. One man could easily do the job.

Q: Can reactors be converted into weapons factories?

A: If a reprocessing plant is available, the plutonium in power reactors is usable for weapons but is of very poor quality for that purpose (cf. Chapter 7). Under nearly all circumstances, a nation desiring nuclear weapons would find it much cheaper, faster, and easier to produce the plutonium in other ways. This would also give their bombs higher explosive power and much improved reliability.

Q: Why did the U.S. Government develop nuclear energy while ignoring solar energy?

A: Any objective analysis would indicate that generating electricity from nuclear energy should be several times cheaper than generating it from solar energy. Moreover, solar energy is only available when the sun shines, which greatly increases the problems and costs if it is used as the principal power source.

No one, at that time, could foresee the political opposition to nuclear power which has driven its cost so high in the United States.

Q: Is the cost of waste disposal and the cost of eventually decommissioning* the plant included in the cost estimates for nuclear power?

A: The cost of waste disposal represents only about 1% of the cost of nuclear electricity. A tax to cover this cost (with something to spare) is included in the waste disposal bill passed by Congress in 1982. The cost of decommissioning is to be borne by the utility, so they include it in their calculation of the rate to charge customers. It contributes only about 1% to that rate.

Q: Your discussion is too technical for us to understand. How can you expect a nonscientist to follow your arguments?

A: I have done my best to make my arguments as understandable as possible. The one thing it is impossible to do, however, is to answer the criticisms of nuclear power without giving *quantitative* demonstrations of the number of deaths that might be expected from nuclear versus alternative technologies. Doing anything less than that, I could easily make any technology appear to be as dangerous, or as safe, as I choose. If you are not willing to follow those quantitative demonstrations, you could just accept the results with the understanding that they have been accepted by the great majority of the scientific community. These results are stated most succinctly in Chapter 4 (results from antinuclear activists in parentheses): having all of our electricity generated by nuclear power plants would reduce our life expectancy by less than one hour (1.5 days), making it as dangerous as a regular smoker smoking one extra cigarette every 15 years (3 months), as an overweight person increasing his weight by 0.012 ounces (0.8 ounces), or as raising the U.S. highway speed limit from 55 to 55.006 (55.4) miles per hour, and it is 2000 times (30 times) less risky than switching from standard size to small cars.

If you are not willing to follow the quantitative risk assessments or to accept their results, you should disqualify yourself from expressing opinions on nuclear power issues. I do not understand how anyone can form an intelligent opinion without giving very heavy weight to these quantitative risk assessments.

---

* Decommissioning refers to taking a plant out of service, dismantling it, and restoring the site for other uses.

Chapter 12 / A CRY FOR HELP

I began this book by showing that there is a vast gulf of misunderstanding of nuclear power issues by the public. Polls show that the public is much more fearful of nuclear power and much less willing to accept it than experts who are in the best position to know the facts. There are a number of reasons for this misunderstanding. Let us review them briefly:

1. Wildly Exaggerated Fear of Radiation

The public views radiation as something very new, highly mysterious, and extremely dangerous. But there is nothing *new* about radiation. Mankind has always been exposed to natural radiation at levels *hundreds of times higher* than it can ever expect to receive from the generation of nuclear electricity. A similar comparison applies to medical X-rays which are a new source of radiation, first introduced in this century, and accepted without fear for at least two generations. Moreover, radiation exposures larger than those from nuclear power are regularly received from airline flights, luminous dial watches, television viewing, the bricks, stones, and plaster from which our buildings are constructed, and other commonly accepted materials and activities. The radiation received from these sources is in no way distinguishable from that of the nuclear industry.

There is also nothing mysterious about radiation, at least from the scientist's viewpoint. It is a relatively simple physical phenomenon, whose health effects are far better understood than those of air pollution, food additives, chemicals, or almost any other environmental threat. It is readily amenable to scientific study, and it has been the subject of extensive scientific investigation for over 40 years. Contrary to the impression created by the media, there is little disagreement within the scientific community over the health effects of radiation.

How dangerous is radiation? That is a quantitative question, and any attempt to treat it otherwise is bound to lead to confusion. In all of the incidents to date that have received so much media coverage, those exposed have less than one chance in a million of having their health affected. In fact they received far less radiation than is given by a single medical X-ray, and no more than what each of us gets every few days from natural sources. The reason this low-level radiation is not very dangerous is *not* because it takes some minimum amount of radiation to harm us—even a single particle of radiation can cause cancer (as can a single molecule of many chemicals). Rather it is because the *probability* for it to cause cancer is extremely small, something like one chance in 50 quadrillion for each particle that strikes us. That is why we are not at great risk even from the 15,000 particles that strike us every second—500 billion per year—from natural sources. It is important to understand that this game of chance with radiation is only one of many thousands of fatal games of chance in which we are, of necessity, engaged. Every breath of air we inhale, every bite of food we swallow, anything or any person we touch, any puff of wind that strikes us, every step we take, every moment we stand still, can bring with it a molecule of material, a micro-organism, or any number of other things ranging in force up to a bolt of lightning that will be fatal to us. The only thing that saves us is that the odds in these games of chance are heavily in our favor. In the case of radiation the odds are more favorable than usual—one chance in 50 quadrillion is pretty good odds.

As a result of inordinately excessive and deceptive media coverage, the public has been driven *insane* over fear of radiation; that is, its perception of the dangers has almost completely lost contact with reality. It has lost all perspective on these dangers relative to those of the many other risks we face.

2. Highly Distorted Pictures of a Reactor Meltdown Accident

The public has been led to believe that the incident at Three Mile Island was a close call to a public health disaster, whereas all postaccident studies agree that it was not. The public has been led to believe that a reactor meltdown is the ultimate disaster, whereas scientific analyses conclude that most meltdowns would not cause even a single death. It does not realize that for nuclear power to be as dangerous as coal burning, our principal present means of generating electricity, there would have to be a reactor meltdown every two weeks somewhere in the United States. It does not appreciate the numerous engineered safety features, the defense in depth, that make a meltdown exceedingly improbable, in spite of various possible equipment malfunctions, human errors, and what have you. It does not appreciate the great added security provided by the containment, which nearly always protects the public regardless of what happens inside. It has been misled, again by excessive media coverage and sensationalism, into believing that reactor meltdowns are one of the major threats we face, whereas even if we accept the estimates of Union of Concerned Scientists, the nation's leading antinuclear activist organization, it is many times less of a threat than such mundane risks as air pollution, falling, drowning while taking a bath, being electrocuted or poisoned or burned, and so forth.

Unfortunately, to understand the intricacies of reactor accidents and their potential consequences requires careful consideration of technical details. In fact, there has even been a considerable confusion over what is, and is not, a safety issue. Chapter 3 represents an attempt to resolve these questions and present the relevant information.

3. Failure to Understand and Quantify Risk

According to most scientific analyses, the loss of life expectancy, LLE, by the average American due to nuclear power, if all of the electricity now used in the United States were generated by that technology, would be only about one hour; even if we accept the estimates of the antinuclear activist Union of Concerned Scientists, it would still be only 1.5 days. By contrast, the LLE due to some other risks are in the range of several *years,* including smoking cigarettes, being unmarried, being poor and/or uneducated, being 10% or more overweight, working as a coal miner, policeman, or shipworker, and living in the Southeastern section of the United States. Each of these risks is at least 10,000 times higher than the risk of nuclear power. Our LLE is in the range of 40–100 days from such risks as using compact cars, having an occupational accident in an average job, and being murdered; each of these is at least a thousand times more of a danger to us than living with nuclear power. Our LLE from air pollution due to coal burning, the principal alternative to nuclear power, is 13 days, and from electrocution, another phase of the electrical industry, is 5 days. With any perspective, therefore, our 1 hour LLE from nuclear power is truly trivial.

Another way of expressing this is to list some risks that are equivalent to the risks to an average American from an all-nuclear electric power system. The numbers given are typical scientific estimates, with Union of Concerned Scientists' estimates in parentheses. Some of these equivalent risks are a regular smoker's indulging in one extra cigarette every 15 years (every 3 months), an overweight person's increasing his weight by 0.01 ounce (0.8 ounces), or raising the U.S. highway speed limit from 55 miles per hour to 55.006 (55.4).

Our present practice of generating electricity by coal burning is 300 (10) times more dangerous than generating it from nuclear fuels. Every time a coal-fired plant is built instead of a nuclear plant, many hundreds of our citizens are unnecessarily condemned to an early death. This is true even if we accept the Union of Concerned Scientists' estimates of the risks from nuclear power.

Other energy sources like oil, gas, solar, and hydroelectric, are also more harmful to public health than nuclear. But the most harmful to health of all energy strategies is overzealous conservation—trying to reduce our energy use by measures going beyond simply eliminating waste.

There are some tragic consequences of the public's insane fear of radiation. Many thousands of lives could be saved each year at costs below $100,000 per life saved by expenditures on health care and highway safety, whereas we are spending tens of millions of dollars per life saved in protecting people against radiation from nuclear power. This is a natural consequence of our government's duty to respond to public concern. The problem is that the public's concern is driven by media coverage rather than by valid scientific information. Thus the media's practice of greatly overemphasizing nuclear power risks while underemphasizing the more important risks is costing our nation thousands of lives unnecessarily lost, and billions of dollars wasted every year.

4. Grossly Unjustified Fears about Disposal of Radioactive Waste

What are we going to do with our radioactive waste? We're simply going to convert it into rocks, and put it in the natural habitat of rocks, deep underground. Will that be safe? If the waste-converted-to-rock behaves like other rock, it will be extremely safe, thousands of times less damaging to human health than the wastes from coal burning, for example. The validity of this comparison between buried waste and other rock has been investigated and found to be well justified.

There are many other comparisons that give proper perspective on the dangers of radioactive waste. It will cause thousands of times fewer fatalities than air pollution due to coal burning, one of the principal wastes from that technology. Some of the materials released in coal burning which will last forever—beryllium, cadmium, arsenic, nickel, and chromium—will eventually cause thousands of times more cancer than the nuclear waste produced in generating the same quantity of electricity. The wastes generated in producing and deploying solar energy systems will kill hundreds of times as many people as those from nuclear wastes.

Problems associated with radon, a naturally radioactive gas evolving from the radioactive decay of uranium, give us other perspectives on the problem. Tightening up our buildings to reduce heat leakage will trap radon inside, thereby causing hundreds of times as many radiation deaths as nuclear power. This makes energy *conservation* by far the most dangerous energy strategy if radiation is what we are worried about. Uranium is present as an impurity in coal and when the coal is burned, it finds its way by one route or another into the ground, serving as a source of radon. By this process alone, coal-burning waste is killing thousands of times as many people as the wastes from nuclear power. There are also radon-emitting wastes from nuclear power in the mill tailings and these, even with careful management, will kill about 30 times as many people as the high-level waste; thus even if we were to consider only the nuclear wastes, we should be worrying about the radon from mill tailings rather than about the high-level waste. But the latter gets all the media coverage, and therefore the vast majority of the public concern, and consequently nearly all of the government attention and money.

But by far the most important health effect of radon caused by the nuclear power industry is the lives *saved* by digging uranium out of the ground, thereby eliminating it as a source of future radon exposure. That process saves tens of thousands of lives for every life lost due to the high-level waste.

Low-level radioactive waste and transuranic waste from the nuclear industry are even less of a hazard than the high-level waste. Transportation hazards associated with radioactive waste are of trivial importance. For example, the number of lives lost due to radioactivity releases in the transport of nuclear waste is estimated to be 1/100,000 of the number we lose from transporting coal. From all of the accidents that have occurred in nuclear waste transport up to the present time, there is less than a 1% chance that there will ever be a single fatality.

Some of the political and media activities in the area of radioactive waste have grossly deceived and misled the public, driving the government into actions that represent a shameful waste of taxpayers' money. Therein lies our nation's real waste problem.

5. Fears of Plutonium and Connections Between Nuclear Power and Nuclear Bombs

Plutonium is often considered to be a curse on mankind, but its use in breeder reactors can supply all the energy mankind will ever need with no increase in cost because of fuel scarcity. In the future it may therefore be considered one of God's greatest gifts to mankind.

Much has been made of the connection between nuclear power and proliferation of nuclear weapons, but on close examination, we find this link to be a very weak one. Only under most unusual circumstances would it be useful for a nation desiring to develop a nuclear weapons capability to use nuclear power plants for that purpose; ordinarily there are much easier, cheaper, and faster methods, which produce more destructive and more reliable bombs. This issue, of course, has no relevance to the use of nuclear power in the United States.

The idea that terrorists might steal plutonium to make nuclear bombs is also a highly exaggerated threat. Terrorists would face truly formidable problems in trying to steal plutonium, and again in trying to fashion it into a bomb. Moreover, the crude nuclear bombs they might make would give them no destructive capabilities they do not already possess through much simpler and more mundane means.

Hazards from plutonium toxicity have also been highly exaggerated. Contrary to widely publicized statements, plutonium is *not* the most toxic substance known to man, a single particle inhaled into the lung is *not* fatal, and the other wild exaggerations propagated by the media are equally misguided. This is another example of the false propaganda that the public has been fed on nuclear power issues.

6. The Romantic Notion that Solar Electricity Could or Should Replace Nuclear Energy or Coal

There are two problems with solar electricity: (1) it is and must always be very expensive compared to what we now pay for electricity, and (2) its availability varies widely between day and night, between summer and winter, and between cloudy and clear days. The long-term goals of the Government's solar photovoltaic development program are to cover large areas of the ground with highly sophisticated electronic devices, each one inch in diameter, that must withstand the outside weather for 20–30 years, at the same price we now pay to cover the same area with cement sidewalks. Even if this miracle is accomplished, the electricity will cost about five times as much as we now pay for electricity.

The variability of solar energy will be even more difficult to live with if we intend to derive all our electricity by that technology. Just to store enough electricity during a sunny day to use that night would cost an average household about a thousand dollars per year, more than the total it now spends for electricity. It would be several times more expensive to provide service during a string of successive cloudy days in winter.

Because of these problems, solar electricity experts consider that technology to be potentially viable only as a supplement for electricity generated by nuclear power or coal burning. But the public does not recognize this. Again due to media distortion and misleading propaganda, plus a heavy dose of wishful thinking, the public has been convinced that we don't need nuclear power or coal, but can depend on "free" energy from the Sun.

Not only is solar electricity extremely expensive, but it is not pollution free. In fact, it is substantially more hazardous to our health than nuclear energy. Here again the public has been deceived.

As a consequence of these misunderstandings, pressure from the public has forced the Nuclear Regulatory Commission into the practice of regulatory ratcheting which has increased the cost of nuclear power by a large factor. As a result, it has now become virtually impossible for a utility to order a new nuclear power plant, and the cost of plants now under construction has skyrocketed.

We are not suffering much from this insanity now, but in the not too distant future, the cost of electricity in the United States will be double what it is now in constant dollars. It will also be nearly double the cost of electricity in Western Europe or Japan. This represents a reversal of the historical situation where U.S. electricity was among the world's cheapest. The effects on our economy are bound to be extremely serious.

In summary, the consequences of the public misunderstandings we have discussed in this book are tragic in the extreme. They are costing our nation thousands of lives and billions of dollars every year, and might easily lead to our economic ruin, which in turn would substantially multiply the human costs.

Who is responsible for this tragedy? The responsibility for the misunderstanding must lie in the source of the public's information—the media.

They have distorted their coverage of nuclear power unmercifully. In some cases their behavior has been unconscionable, but more often it has been driven by the fact that they are basically in the entertainment business. Educating the public must take a back seat in their priorities to providing the interest and entertainment needed to grab and hold their audience. There is no other way a media outlet can survive in such a competitive industry.

Thus the ultimate victim and culprit in this tragedy is the American public. By requiring that their information be fed to them in an entertaining format, by allowing their thinking about scientific issues to be guided by media presentations rather than asking to hear the scientists' viewpoint, and by exercising their democratic rights to form public policy on scientific and technological issues without fulfilling the concomitant duty of obtaining the quantitative information needed to make those decisions, the American public is bringing this tragedy upon itself.

But more relevant than assigning blame is the question of what can be done to avert, or at least to mitigate this tragedy. There is little in my education or experience that qualifies me as an expert, or even as a competent practitioner in that area. Writing this book is my cry for help from any reader who may have the needed expertise. We in the scientific community stand ready to help in every way possible. We will provide information, we will transform it into any form desired, we will speak or write for audiences that can understand us.

Most of us are willing, indeed eager, to devote our time and energy to these activities without material compensation. Our only goal is to mitigate the suffering of the American people from the needless tragedy that threatens to engulf them. We will do anything we can to strive for this goal.

If, with your help, we can clear up the public misunderstandings and turn the situation around, we may still enjoy the wonderful blessings that nuclear power is so capable of bringing us—cheap and abundant energy forever, along with substantially improved health, safety, and preservation of our environment.

INDEX